Curved Spaces

This self-contained textbook presents an expositi ... ...o-
dimensional geometries, such as Euclidean, spherical, hyperbolic and the locally Euclidean torus, and introduces the basic concepts of Euler numbers for topological triangulations and Riemannian metrics. The careful discussion of these classical examples provides students with an introduction to the more general theory of curved spaces developed later in the book, as represented by embedded surfaces in Euclidean 3-space, and their generalization to abstract surfaces equipped with Riemannian metrics. Themes running throughout include those of geodesic curves, polygonal approximations to triangulations, Gaussian curvature, and the link to topology provided by the Gauss–Bonnet theorem.

Numerous diagrams help bring the key points to life and helpful examples and exercises are included to aid understanding. Throughout the emphasis is placed on explicit proofs, making this text ideal for any student with a basic background in analysis and algebra.

PELHAM WILSON is Professor of Algebraic Geometry in the Department of Pure Mathematics, University of Cambridge. He has been a Fellow of Trinity College since 1981 and has held visiting positions at universities and research institutes worldwide, including Kyoto University and the Max-Planck-Institute for Mathematics in Bonn. Professor Wilson has over 30 years of extensive experience of undergraduate teaching in mathematics, and his research interests include complex algebraic varieties, Calabi–Yau threefolds, mirror symmetry and special Lagrangian submanifolds.

Curved Spaces

From Classical Geometries to Elementary Differential Geometry

P. M. H. Wilson

Department of Pure Mathematics, University of Cambridge,
and Trinity College, Cambridge

CAMBRIDGE
UNIVERSITY PRESS

CAMBRIDGE UNIVERSITY PRESS
Cambridge, New York, Melbourne, Madrid, Cape Town, Singapore, São Paulo

Cambridge University Press
The Edinburgh Building, Cambridge CB2 8RU, UK

Published in the United States of America by Cambridge University Press, New York

www.cambridge.org
Information on this title: www.cambridge.org/9780521886291

First published 2008

Printed in the United Kingdom at the University Press, Cambridge

A catalogue record for this publication is available from the British Library

ISBN 978-0-521-88629-1 hardback
ISBN 978-0-521-71390-0 paperback

Cambridge University Press has no responsibility for the persistence or accuracy
of URLs for external or third-party internet websites referred to in this publication,
and does not guarantee that any content on such websites is, or will remain,
accurate or appropriate.

For Stanzi, Toby and Alexia,
in the hope that one day
they might understand what is written herein,
and to Sibylle.

Contents

Preface

This book represents an expansion of the author's lecture notes for a course in Geometry, given in the second year of the Cambridge Mathematical Tripos. Geometry tends to be a neglected part of many undergraduate mathematics courses, despite the recent history of both mathematics and theoretical physics being marked by the continuing importance of geometrical ideas. When an undergraduate geometry course is given, it is often in a form which covers various assorted topics, without necessarily having an underlying theme or philosophy — the author has in the past given such courses himself. One of the aims in this volume has been to set the well-known classical two-dimensional geometries, Euclidean, spherical and hyperbolic, in a more general context, so that certain geometrical themes run throughout the book. The geometries come equipped with well-behaved distance functions, which in turn give rise to curvature of the space. The curved spaces in the title of this book will nearly always be two-dimensional, but this still enables us to study such basic geometrical ideas as geodesics, curvature and topology, and to understand how these ideas are interlinked. The classical examples will act both as an introduction to, and examples of, the more general theory of curved spaces studied later in the book, as represented by embedded surfaces in Euclidean 3-space, and more generally by abstract surfaces with Riemannian metrics.

The author has tried to make this text as self-contained as possible, although the reader will find it very helpful to have been exposed to first courses in Analysis, Algebra, and Complex Variables beforehand. The course is intended to act as a link between these basic undergraduate courses, and more theoretical geometrical theories, as represented say by courses on Riemann Surfaces, Differential Manifolds, Algebraic Topology or Riemannian Geometry. As such, the book is not intended to be another text on Differential Geometry, of which there are many good ones in the literature, but has rather different aims. For books on differential geometry, the author can recommend three in particular, which he has consulted when writing this volume, namely [5], [8] and [9]. The author has also not attempted to put the geometry he describes into a historical perspective, as for instance is done in [8].

As well as making the text as self-contained as possible, the author has tried to make it as elementary and as explicit as possible, where the use of the word elementary here implies that we wish to rely as little as possible on theory developed elsewhere. This explicit approach does result in one proof where the general argument is both intuitive and clear, but where the specific details need care to get correct, the resulting formal proof therefore being a little long. This proof has been placed in an appendix to Chapter 3, and the reader wishing to maintain his or her momentum should skip over this on first reading. It may however be of interest to work through this proof at some stage, as it is by understanding where the problems lie that the more theoretical approach will subsequently be better appreciated. The format of the book has however allowed the author to be more expansive than was possible in the lectured course on certain other topics, including the important concepts of differentials and abstract surfaces. It is hoped that the latter parts of the book will also serve as a useful resource for more advanced courses in differential geometry, where our concrete approach will complement the usual rather more abstract treatments.

The author wishes to thank Nigel Hitchin for showing him the lecture notes of a course on Geometry of Surfaces he gave in Oxford (and previously given by Graeme Segal), which will doubtless have influenced the presentation that has been given here. He is grateful to Gabriel Paternain, Imre Leader and Dan Jane for their detailed and helpful comments concerning the exposition of the material, and to Sebastian Pancratz for his help with the diagrams and typesetting. Most importantly, he wishes to thank warmly his colleague Gabriel Paternain for the benefit of many conversations around the subject, which have had a significant impact on the final shape of the book.

1 Euclidean geometry

1.1 Euclidean space

Our story begins with a geometry which will be familiar to all readers, namely the geometry of Euclidean space. In this first chapter we study the Euclidean distance function, the symmetries of Euclidean space and the properties of curves in Euclidean space. We also generalize some of these ideas to the more general context of metric spaces, and we sketch the basic theory of metric spaces, which will be needed throughout the book.

We consider Euclidean space $\mathbf{R}^n$, equipped with the standard Euclidean inner-product $(\,,\,)$, which we also refer to as the dot product; namely, given vectors $\mathbf{x}, \mathbf{y} \in \mathbf{R}^n$ with coordinates x_i, y_i respectively, the inner-product is defined by

$$(\mathbf{x}, \mathbf{y}) = \sum_{i=1}^{n} x_i y_i.$$

We then have a *Euclidean norm* on $\mathbf{R}^n$ defined by $\|\mathbf{x}\| = (\mathbf{x}, \mathbf{x})^{1/2}$, and a distance function d defined by

$$d(\mathbf{x}, \mathbf{y}) = \|\mathbf{x} - \mathbf{y}\|.$$

In some books, the Euclidean space will be denoted $\mathbf{E}^n$ to distinguish it from the vector space $\mathbf{R}^n$, but we shall not make this notational distinction.

The Euclidean distance function d is an example of a *metric*, in that for any points P, Q, R of the space, the following three conditions are satisfied:

(i) $d(P, Q) \geq 0$, with equality if and only if $P = Q$.
(ii) $d(P, Q) = d(Q, P)$.
(iii) $d(P, Q) + d(Q, R) \geq d(P, R)$.

The crucial condition here is the third one, which is known as the *triangle inequality*. In the Euclidean case, it says that, for a Euclidean triangle (possibly degenerate) with vertices P, Q and R, the sum of the lengths of two sides of the triangle is at least the length of the third side. In other words, if one travels (along straight line segments) from P to R via Q, the length of one's journey is at least that of the direct route from P to R.

To prove the triangle equality in the Euclidean case, we use the Cauchy–Schwarz inequality, namely

$$\left(\sum_{i=1}^{n} x_i y_i\right)^2 \leq \left(\sum_{i=1}^{n} x_i^2\right)\left(\sum_{i=1}^{n} y_i^2\right),$$

or, in the inner-product notation, that $(\mathbf{x}, \mathbf{y})^2 \leq \|x\|^2 \|y\|^2$. The Cauchy–Schwarz inequality also includes the criterion for equality to hold, namely that the vectors $\mathbf{x}$ and $\mathbf{y}$ should be proportional. We may prove the Cauchy–Schwarz inequality directly from the fact that, for any $\mathbf{x}, \mathbf{y} \in \mathbf{R}^n$, the quadratic polynomial in the real variable λ,

$$(\lambda \mathbf{x} + \mathbf{y}, \lambda \mathbf{x} + \mathbf{y}) = \|x\|^2 \lambda^2 + 2(\mathbf{x}, \mathbf{y})\lambda + \|y\|^2,$$

is positive semi-definite. Furthermore, equality holds in Cauchy–Schwarz if and only if the above quadratic polynomial is indefinite; assuming $\mathbf{x} \neq \mathbf{0}$, this just says that for some $\lambda \in \mathbf{R}$, we have $(\lambda \mathbf{x} + \mathbf{y}, \lambda \mathbf{x} + \mathbf{y}) = 0$, or equivalently that $\lambda \mathbf{x} + \mathbf{y} = \mathbf{0}$.

To see that the triangle inequality follows from the Cauchy–Schwarz inequality, we may take P to be the origin in $\mathbf{R}^n$, the point Q to have position vector $\mathbf{x}$ with respect to the origin, and R to have position vector $\mathbf{y}$ with respect to Q, and hence position vector $\mathbf{x} + \mathbf{y}$ with respect to the origin. The triangle inequality therefore states that

$$(\mathbf{x} + \mathbf{y}, \mathbf{x} + \mathbf{y})^{1/2} \leq \|\mathbf{x}\| + \|\mathbf{y}\|;$$

on squaring and expanding, this is seen to be equivalent to the Cauchy–Schwarz inequality.

In the Euclidean case, we have a characterization for equality to hold; if it does, then we must have equality holding in the Cauchy–Schwarz inequality, and hence that $\mathbf{y} = \lambda \mathbf{x}$ for some $\lambda \in \mathbf{R}$ (assuming $\mathbf{x} \neq \mathbf{0}$). Equality then holds in the triangle inequality if and only if $|\lambda + 1| \|x\| = (|\lambda| + 1) \|x\|$, or equivalently that $\lambda \geq 0$. In summary therefore, we have equality in the triangle inequality if and only if Q is on the straight line segment PR, in which case the direct route from P to R automatically passes through Q. Most of the metrics we encounter in this course will have an analogous such characterization of equality.

Definition 1.1   A *metric space* is a set X equipped with a *metric d*, namely a function $d : X \times X \to \mathbf{R}$ satisfying the above three conditions.

The basic theory of metric spaces is covered well in a number of elementary textbooks, such as [13], and will be known to many readers. We have seen above that Euclidean space of dimension n forms a metric space; for an arbitrary metric space (X, d), we can generalize familiar concepts from Euclidean space, such as:

- $B(P, \delta) := \{Q \in X : d(Q, P) < \delta\}$, *the open ball of radius δ around a point P.*
- *open sets U in X*: by definition, for each $P \in U$, there exists $\delta > 0$ with $B(P, \delta) \subset U$.
- *closed sets in X*: that is, subsets whose complement in X is open.
- *open neighbourhoods of $P \in X$*: by definition, open sets containing P.

Given two metric spaces (X, d_X), (Y, d_Y), and a function $f : X \to Y$, the usual definition of continuity also holds. We say that f is *continuous* at $P \in X$ if, for any $\varepsilon > 0$, there exists $\delta > 0$ such that $d_X(Q, P) < \delta$ implies that $d_Y(f(Q), f(P)) < \varepsilon$. This last statement may be reinterpreted as saying that the inverse image of $B(f(P), \varepsilon)$ under f contains $B(P, \delta)$.

Lemma 1.2   *A map $f : X \to Y$ of metric spaces is continuous if and only if, under f, the inverse image of every open subset of Y is open in X.*

Proof   If f is continuous, and U is an open subset of Y, we consider an arbitrary point $P \in f^{-1}U$. Since $f(P) \in U$, there exists $\varepsilon > 0$ such that $B(f(P), \varepsilon) \subset U$. By continuity, there exists an open ball $B(P, \delta)$ contained in $f^{-1}(B(f(P), \varepsilon)) \subset f^{-1}U$. Since this holds for all $P \in f^{-1}U$, it follows that $f^{-1}U$ is open.

Conversely, suppose now that this condition holds for all open sets U of Y. Given any $P \in X$ and $\varepsilon > 0$, we have that $f^{-1}(B(f(P), \varepsilon))$ is an open neighbourhood of P, and hence there exists $\delta > 0$ with $B(P, \delta) \subset f^{-1}(B(f(P), \varepsilon))$. $\square$

Thus, continuity of f may be phrased purely in terms of the open subsets of X and Y. We say therefore that continuity is defined *topologically*.

Given metric spaces (X, d_X) and (Y, d_Y), a *homeomorphism* between them is just a continuous map with a continuous inverse. By Lemma 1.2, this is saying that the open sets in the two spaces correspond under the bijection, and hence that the map is a *topological* equivalence between the spaces; the two spaces are then said to be *homeomorphic*. Thus for instance, the open unit disc $D \subset \mathbf{R}^2$ is homeomorphic to the whole plane (both spaces with the Euclidean metric) via the map $f : D \to \mathbf{R}^2$ given by $f(\mathbf{x}) = \mathbf{x}/(1 - \|\mathbf{x}\|)$, with inverse $g : \mathbf{R}^2 \to D$ given by $g(\mathbf{y}) = \mathbf{y}/(1 + \|\mathbf{y}\|)$.

All the geometries studied in this book will have natural underlying metric spaces. These metric spaces will however have particularly nice properties; in particular they have the property that every point has an open neighbourhood which is homeomorphic to the open disc in $\mathbf{R}^2$ (this is essentially the statement that the metric space is what is called a two-dimensional *topological manifold*). We conclude this section by giving two examples of metric spaces, both of which are defined geometrically but neither of which have this last property.

Example (British Rail metric)   Consider the plane $\mathbf{R}^2$ with Euclidean metric d, and let O denote the origin. We define a new metric d_1 on $\mathbf{R}^2$ by

$$d_1(P, Q) = \begin{cases} 0 & \text{if } P = Q, \\ d(P, O) + d(O, Q) & \text{if } P \neq Q. \end{cases}$$

We note that, for $P \neq O$, any small enough open ball round P consists of just the point P; therefore, no open neighbourhood of P is homeomorphic to an open disc in $\mathbf{R}^2$. When the author was an undergraduate in the UK, this was known as the *British Rail* metric; here O represented London, and all train journeys were forced to go via London! Because of a subsequent privatization of the UK rail network, the metric should perhaps be renamed.

Example (London Underground metric)   Starting again with the Euclidean plane $(\mathbf{R}^2, d)$, we choose a finite set of points $P_1, \ldots, P_N \in \mathbf{R}^2$. Given two points $P, Q \in \mathbf{R}^2$, we define a distance function d_2 (for $N > 1$, it is not a metric) by

$$d_2(P, Q) = \min\{d(P, Q), \min_{i,j}\{d(P, P_i) + d(P_j, Q)\}\}.$$

This function satisfies all the properties of a metric *except* that $d_2(P, Q)$ may be zero even when $P \neq Q$. We can however form a quotient set X from $\mathbf{R}^2$ by identifying all the points P_i to a single point $\bar{P}$ (formally, we take the quotient of $\mathbf{R}^2$ by the equivalence relation which sets two points P, Q to be equivalent if and only if $d_2(P, Q) = 0$), and it is then easily checked that d_2 induces a metric d^* on X. The name given to this metric refers to the underground railway in London; the points P_i represent the idealized stations in this network, idealized because we assume that no walking is involved if we wish to travel between any two stations of the network (even if such a journey involves changing trains). The distance d_2 between two points of $\mathbf{R}^2$ is the minimum distance one has to walk between the two points, given that one has the option of walking to the nearest underground station and travelling by train to the station nearest to one's destination.

We note that any open ball of sufficiently small radius ε round the point $\bar{P}$ of X corresponding to the points $P_1, \ldots, P_N \in \mathbf{R}^2$ is the union of the open balls $B(P_i, \varepsilon) \subset \mathbf{R}^2$, with the points $P_1, \ldots, P_N$ identified. In particular, the punctured ball $B(\bar{P}, \varepsilon) \backslash \{\bar{P}\}$ in X is identified as a disjoint union of punctured balls $B(P_i, \varepsilon) \backslash \{P_i\}$ in the plane. Once we have introduced the concept of connectedness in Section 1.4, it will be clear that this latter space is not connected for $N \geq 2$, and hence cannot be homeomorphic to an open punctured disc in $\mathbf{R}^2$, which from Section 1.4 is clearly both connected and path connected. It will follow then that our open ball in X is not homeomorphic to an open disc in $\mathbf{R}^2$. The same is true for any open neighbourhood of $\bar{P}$.

1.2   Isometries

We defined above the concept of a *homeomorphism* or *topological equivalence*; the geometries in this course however come equipped with metrics, and so we shall be interested in the stronger notion of an *isometry*.

Definition 1.3   A map $f : (X, d_X) \to (Y, d_Y)$ between metric spaces is called an *isometry* if it is surjective and it preserves distances, that is

$$d_Y(f(x_1), f(x_2)) = d_X(x_1, x_2)$$

for all $x_1, x_2 \in X$.

A few observations are due here:

- The second condition in (1.3) implies that the map is injective. Thus an isometry is necessarily bijective. A map satisfying the second condition without necessarily being surjective is usually called an *isometric embedding*.
- The second condition implies that an isometry is continuous, as is its inverse. Hence isometries are homeomorphisms. However, the homeomorphism defined above between the unit disc and the Euclidean plane is clearly not an isometry.
- An isometry of a metric space to itself is also called a *symmetry* of the space. The isometries of a metric space X to itself form a group under composition of maps, called the *isometry group* or the *symmetry group* of the space, denoted Isom(X).

Definition 1.4   We say that a group G *acts* on a set X if there is a map $G \times X \to X$, the image of (g, x) being denoted by $g(x)$, such that

(i)   the identity element in G corresponds to the identity map on X, and

(ii) $(g_1 g_2)(x) = g_1(g_2(x))$ for all $x \in X$ and $g_1, g_2 \in G$.

We say that the action of G is *transitive* on X if, for all $x, y \in X$, there exists $g \in G$ with $g(x) = y$.

For X a metric space, the obvious action of Isom(X) on X will not usually be transitive. For the important special cases however of Euclidean space, the sphere (Chapter 2), the locally Euclidean torus (Chapter 3) and the hyperbolic plane (Chapter 5), this action is transitive — these geometries may therefore be thought of as looking the same from every point.

Let us now consider the case of Euclidean space $\mathbf{R}^n$, with its standard inner-product $(\,,\,)$ and distance function d. An isometry of $\mathbf{R}^n$ is sometimes called a *rigid motion*. We note that any translation of $\mathbf{R}^n$ is an isometry, and hence the isometry group Isom($\mathbf{R}^n$) acts transitively on $\mathbf{R}^n$.

We recall that an $n \times n$ matrix A is called *orthogonal* if $A^t A = A A^t = I$, where A^t denotes the transposed matrix. Since

$$(A\mathbf{x}, A\mathbf{y}) = (A\mathbf{x})^t (A\mathbf{y})$$
$$= \mathbf{x}^t A^t A\mathbf{y}$$
$$= (\mathbf{x}, A^t A\mathbf{y})$$
$$= (A^t A\mathbf{x}, \mathbf{y}),$$

we have that A is orthogonal if and only if $(A\mathbf{x}, A\mathbf{y}) = (\mathbf{x}, \mathbf{y})$ for all $\mathbf{x}, \mathbf{y} \in \mathbf{R}^n$.

Since $(\mathbf{x}, \mathbf{y}) = \frac{1}{2}\{\|\mathbf{x} + \mathbf{y}\|^2 - \|\mathbf{x}\|^2 - \|\mathbf{y}\|^2\}$, a matrix A is orthogonal if and only if $\|A\mathbf{x}\| = \|\mathbf{x}\|$ for all $\mathbf{x} \in \mathbf{R}^n$. Thus, if a map $f : \mathbf{R}^n \to \mathbf{R}^n$ is defined by $f(\mathbf{x}) = A\mathbf{x} + \mathbf{b}$, for some $\mathbf{b} \in \mathbf{R}^n$, then

$$d(f(\mathbf{x}), f(\mathbf{y})) = \|A(\mathbf{x} - \mathbf{y})\|,$$

and so f is an isometry if and only if A is orthogonal.

Theorem 1.5   *Any isometry* $f : \mathbf{R}^n \to \mathbf{R}^n$ *is of the form* $f(\mathbf{x}) = A\mathbf{x} + \mathbf{b}$, *for some orthogonal matrix A and vector $\mathbf{b} \in \mathbf{R}^n$.*

Proof   Let $e_1, \ldots, e_n$ denote the standard basis of $\mathbf{R}^n$. We set $\mathbf{b} = f(\mathbf{0})$, and $\mathbf{a}_i = f(e_i) - \mathbf{b}$ for $i = 1, \ldots, n$. Then, for all i,

$$\|\mathbf{a}_i\| = \|f(e_i) - f(\mathbf{0})\| = d(f(e_i), f(\mathbf{0}))$$
$$= d(e_i, \mathbf{0}) = \|e_i\| = 1.$$

For $i \neq j$,

$$(\mathbf{a}_i, \mathbf{a}_j) = -\frac{1}{2}\left\{\|\mathbf{a}_i - \mathbf{a}_j\|^2 - \|\mathbf{a}_i\|^2 - \|\mathbf{a}_j\|^2\right\}$$
$$= -\frac{1}{2}\left\{\|f(e_i) - f(e_j)\|^2 - 2\right\}$$
$$= -\frac{1}{2}\left\{\|e_i - e_j\|^2 - 2\right\} = 0.$$

$\square$

Now let A be the matrix with columns $\mathbf{a}_1, \ldots, \mathbf{a}_n$. Since the columns form an orthonormal basis, A is orthogonal. Let $g : \mathbf{R}^n \to \mathbf{R}^n$ be the isometry given by

$$g(\mathbf{x}) = A\mathbf{x} + \mathbf{b}.$$

Then $g(\mathbf{x}) = f(\mathbf{x})$ for $\mathbf{x} = \mathbf{0}, e_1, \ldots, e_n$. Now g has inverse g^{-1}, where

$$g^{-1}(\mathbf{x}) = A^{-1}(\mathbf{x} - \mathbf{b}) = A^t(\mathbf{x} - \mathbf{b});$$

therefore $h = g^{-1} \circ f$ is an isometry fixing $\mathbf{0}, e_1, \ldots, e_n$.
   We claim that $h = \mathrm{id}$, and hence $f = g$ as required.

Claim $h : \mathbf{R}^n \to \mathbf{R}^n$ is the identity map.

Proof of Claim   For general $\mathbf{x} = \sum x_i e_i$, set

$$h(\mathbf{x}) = \mathbf{y} = \sum y_i e_i.$$

We observe that

$$d(\mathbf{x}, e_i)^2 = \|\mathbf{x}\|^2 + 1 - 2x_i$$
$$d(\mathbf{x}, \mathbf{0})^2 = \|\mathbf{x}\|^2$$
$$d(\mathbf{y}, e_i)^2 = \|\mathbf{y}\|^2 + 1 - 2y_i$$
$$d(\mathbf{y}, \mathbf{0})^2 = \|\mathbf{y}\|^2.$$

Since h is an isometry such that $h(\mathbf{0}) = \mathbf{0}$, $h(e_i) = e_i$ and $h(\mathbf{x}) = \mathbf{y}$, we deduce that $\|\mathbf{y}\|^2 = \|\mathbf{x}\|^2$ and $x_i = y_i$ for all i, i.e. $h(\mathbf{x}) = \mathbf{y} = \sum x_i e_i = \mathbf{x}$ for all $\mathbf{x}$. Thus $h = \mathrm{id}$.

$\square$

Example (Reflections in affine hyperplanes)   If $H \subset \mathbf{R}^n$ is an affine hyperplane defined by

$$\mathbf{u} \cdot \mathbf{x} = c$$

for some unit vector $\mathbf{u}$ and constant $c \in \mathbf{R}$, we define a map R_H, the *reflection* in H, by

$$R_H : \mathbf{x} \mapsto \mathbf{x} - 2(\mathbf{x} \cdot \mathbf{u} - c)\,\mathbf{u}.$$

Note that $R(\mathbf{x}) = \mathbf{x}$ for all $\mathbf{x} \in H$. Moreover, one checks easily that any $\mathbf{x} \in \mathbf{R}^n$ may be written uniquely in the form $\mathbf{a} + t\mathbf{u}$, for some $\mathbf{a} \in H$ and $t \in \mathbf{R}$, and that $R_H(\mathbf{a} + t\mathbf{u}) = \mathbf{a} - t\mathbf{u}$.

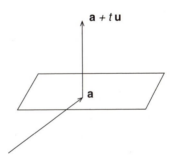

Since

$$(R_H(\mathbf{a} + t\mathbf{u}), R_H(\mathbf{a} + t\mathbf{u})) = (\mathbf{a} - t\mathbf{u}, \mathbf{a} - t\mathbf{u}) = (\mathbf{a}, \mathbf{a}) + t^2 = (\mathbf{a} + t\mathbf{u}, \mathbf{a} + t\mathbf{u}),$$

we deduce that R_H is an isometry.

Conversely, suppose that S is an isometry fixing H (pointwise) and choose any $\mathbf{a} \in H$; if $T_{\mathbf{a}}$ denotes translation by $\mathbf{a}$, i.e. $T_{\mathbf{a}}(\mathbf{x}) = \mathbf{x} + \mathbf{a}$ for all $\mathbf{x}$, then the conjugate $R = T_{-\mathbf{a}}ST_{\mathbf{a}}$ is an isometry fixing pointwise the hyperplane $H' = T_{-\mathbf{a}}H$ through the origin. If H is given by $\mathbf{x} \cdot \mathbf{u} = c$ (where $c = \mathbf{a} \cdot \mathbf{u}$), then H' is given by $\mathbf{x} \cdot \mathbf{u} = 0$. Therefore, $(R\mathbf{u}, \mathbf{x}) = (R\mathbf{u}, R\mathbf{x}) = (\mathbf{u}, \mathbf{x}) = 0$ for all $\mathbf{x} \in H'$, and so $R\mathbf{u} = \lambda\mathbf{u}$ for some λ.

But $\|R\mathbf{u}\|^2 = 1 \implies \lambda^2 = 1 \implies \lambda = \pm 1$. Since, by (1.5), R is a linear map, we deduce that either $R = \mathrm{id}$ or $R = R_{H'}$.

Therefore either $S = \mathrm{id}$, or $S = T_{\mathbf{a}}R_{H'}T_{-\mathbf{a}}$:

$$\mathbf{x} \mapsto \mathbf{x} - \mathbf{a} \mapsto (\mathbf{x} - \mathbf{a}) - 2(\mathbf{x} \cdot \mathbf{u} - \mathbf{a} \cdot \mathbf{u})\mathbf{u} \mapsto \mathbf{x} - 2(\mathbf{x} \cdot \mathbf{u} - c)\mathbf{u},$$

i.e. $S = R_H$.

We shall need the following elementary but useful fact about reflections.

Lemma 1.6　　*Given points $P \neq Q$ in $\mathbf{R}^n$, there exists a hyperplane H, consisting of the points of $\mathbf{R}^n$ which are equidistant from P and Q, for which the reflection R_H swaps the points P and Q.*

Proof　If the points P and Q are represented by vectors $\mathbf{p}$ and $\mathbf{q}$, we consider the perpendicular bisector of the line segment PQ, which is a hyperplane H with equation

$$\mathbf{x} \cdot (\mathbf{p} - \mathbf{q}) = \frac{1}{2}(\|\mathbf{p}\|^2 - \|\mathbf{q}\|^2).$$

An elementary calculation confirms that H consists precisely of the points which are equidistant from P and Q. We observe that $R_H(\mathbf{p} - \mathbf{q}) = -(\mathbf{p} - \mathbf{q})$; moreover $(\mathbf{p}+\mathbf{q})/2 \in H$ and hence is fixed under R_H. Noting that $\mathbf{p} = (\mathbf{p}+\mathbf{q})/2+(\mathbf{p}-\mathbf{q})/2$ and $\mathbf{q} = (\mathbf{p}+\mathbf{q})/2 - (\mathbf{p}-\mathbf{q})/2$, it follows therefore that $R_H(\mathbf{p}) = \mathbf{q}$ and $R_H(\mathbf{q}) = \mathbf{p}$. $\square$

Reflections in hyperplanes form the building blocks for all the isometries, in that they yield generators for the full group of isometries. More precisely, we have the following classical result.

Theorem 1.7　　*Any isometry of $\mathbf{R}^n$ can be written as the composite of at most $(n + 1)$ reflections.*

Proof　As before, we let $e_1, \ldots, e_n$ denote the standard basis of $\mathbf{R}^n$, and we consider the $n+1$ points represented by the vectors $\mathbf{0}, e_1, \ldots, e_n$. Suppose that f is an arbitrary isometry of $\mathbf{R}^n$, and consider the images $f(\mathbf{0}), f(e_1), \ldots, f(e_n)$ of these vectors. If $f(\mathbf{0}) = \mathbf{0}$, we set $f_1 = f$ and proceed to the next step. If not, we use Lemma 1.6; if H_0 denotes the hyperplane of points equidistant from $\mathbf{0}$ and $f(\mathbf{0})$, the reflection R_{H_0} swaps the points. In particular, if we set $f_1 = R_{H_0} \circ f$, then f_1 is an isometry (being the composite of isometries) which fixes $\mathbf{0}$.

We now repeat this argument. Suppose, by induction, that we have an isometry f_i, which is the composite of our original isometry f with at most i reflections, which fixes all the points $\mathbf{0}, e_1, \ldots, e_{i-1}$. If $f_i(e_i) = e_i$, we set $f_{i+1} = f_i$. Otherwise, we let H_i denote the hyperplane consisting of points equidistant from e_i and $f_i(e_i)$. Our assumptions imply that $\mathbf{0}, e_1, \ldots, e_{i-1}$ are equidistant from e_i and $f_i(e_i)$, and hence lie in H_i. Thus R_{H_i} fixes $\mathbf{0}, e_1, \ldots, e_{i-1}$ and swaps e_i and $f_i(e_i)$, and so the composite $f_{i+1} = R_{H_i} \circ f_i$ is an isometry fixing $\mathbf{0}, e_1, \ldots, e_i$.

After $n + 1$ steps, we attain an isometry f_{n+1}, the composite of f with at most $n + 1$ reflections, which fixes all of $\mathbf{0}, e_1, \ldots, e_n$. We saw however in the proof of Theorem 1.5 that this is sufficient to imply that f_{n+1} is the identity, from which it follows that the original isometry f is the composite of at most $n + 1$ reflections. $\square$

Remark　　If we know that an isometry f already fixes the origin, the above proof shows that it can be written as the composite of at most n reflections. The above theorem for $n = 2$, that any isometry of the Euclidean plane may be written as the composite of at most three reflections, has an analogous result in both the spherical and hyperbolic geometries, introduced in later chapters.

1.3   The group $O(3, \mathbf{R})$

A natural subgroup of $\mathrm{Isom}(\mathbf{R}^n)$ consists of those isometries fixing the origin, which can therefore be written as a composite of at most n reflections. By Theorem 1.5, this subgroup may be identified with the group $O(n) = O(n, \mathbf{R})$ of $n \times n$ orthogonal matrices, the *orthogonal group*. If $A \in O(n)$, then

$$\det A \, \det A^t = \det(A)^2 = 1,$$

and so $\det A = \pm 1$. The subgroup of $O(n)$ consisting of elements with $\det A = 1$ is denoted $SO(n)$, and is called the *special orthogonal group*. The isometries f of $\mathbf{R}^n$ of the form $f(\mathbf{x}) = A\mathbf{x} + \mathbf{b}$, for some $A \in SO(n)$ and $\mathbf{b} \in \mathbf{R}^n$, are called the *direct isometries* of $\mathbf{R}^n$; they are the isometries which can be expressed as a product of an *even* number of reflections.

Example   Let us consider the group $O(2)$, which may also be identified as the group of isometries of $\mathbf{R}^2$ fixing the origin. Note that

$$A = \begin{pmatrix} a & b \\ c & d \end{pmatrix} \in O(2) \iff a^2 + c^2 = 1, \, b^2 + d^2 = 1, \, ab + cd = 0.$$

For such a matrix $A \in O(2)$, we may set

$$a = \cos\theta, \qquad c = \sin\theta,$$
$$b = -\sin\phi, \quad d = \cos\phi,$$

with $0 \le \theta, \phi < 2\pi$. Then, the equation $ab + cd = 0$ gives $\tan\theta = \tan\phi$, and therefore $\phi = \theta$ or $\theta \pm \pi$.

In the first case,

$$A = \begin{pmatrix} \cos\theta & -\sin\theta \\ \sin\theta & \cos\theta \end{pmatrix}$$

is an anticlockwise rotation through θ, and $\det A = 1$; it is therefore the product of two reflections. In the second case,

$$A = \begin{pmatrix} \cos\theta & \sin\theta \\ \sin\theta & -\cos\theta \end{pmatrix}$$

is a reflection in the line at angle $\theta/2$ to the x-axis, and $\det A = -1$.

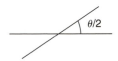

In summary therefore, the elements of $SO(2)$ correspond to the rotations of $\mathbf{R}^2$ about the origin, whilst the elements of $O(2)$ which are not in $SO(2)$ correspond to reflections in a line through the origin.

In this section, we study in more detail the case $n = 3$. We suppose then that $A \in O(3)$. Consider first the case when $A \in SO(3)$, i.e.

$$\boxed{\det A = 1}$$

Then

$$\det(A - I) = \det(A^t - I) = \det A(A^t - I) = \det(I - A)$$
$$\implies \det(A - I) = 0,$$

i.e. $+1$ is an eigenvalue. There exists therefore an eigenvector v_1 (where we may assume $\|v_1\| = 1$) such that $Av_1 = v_1$. Set $W = \langle v_1 \rangle^\perp$ to be the orthogonal complement to the space spanned by v_1. If $w \in W$, then $(Aw, v_1) = (Aw, Av_1) = (w, v_1) = 0$. Thus $A(W) \subset W$ and $A|_W$ is a rotation of the two-dimensional space W, since it is an isometry of W fixing the origin and has determinant one. If $\{v_2, v_3\}$ is an orthonormal basis for W, the action of A on $\mathbf{R}^3$ is represented with respect to the orthonormal basis $\{v_1, v_2, v_3\}$ by the matrix

$$\begin{pmatrix} 1 & 0 & 0 \\ 0 & \cos\theta & -\sin\theta \\ 0 & \sin\theta & \cos\theta \end{pmatrix}.$$

This is just rotation about the axis spanned by v_1 through an angle θ. It may be expressed as a product of two reflections.

Now suppose

$$\boxed{\det A = -1}$$

Using the previous result, there exists an orthonormal basis with respect to which $-A$ is a rotation of the above form, and so A takes the form

$$\begin{pmatrix} -1 & 0 & 0 \\ 0 & \cos\phi & -\sin\phi \\ 0 & \sin\phi & \cos\phi \end{pmatrix}$$

with $\phi = \theta + \pi$. Such a matrix A represents a *rotated reflection*, rotating through an angle ϕ about a given axis and then reflecting in the plane orthogonal to the axis. In the special case $\phi = 0$, A is a pure reflection. The general rotated reflection may be expressed as a product of three reflections.

Example   Consider the rigid motions of $\mathbf{R}^3$ arising from the full symmetry group of a regular tetrahedron **T**, centred on the origin.

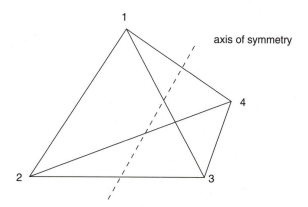

It is clear that the full symmetry group of **T** is S_4, the symmetric group on the four vertices, and that the rotation group of **T** is A_4. Apart from the identity, the rotations either have an axis passing through a vertex and the midpoint of an opposite face, the angle of rotation being $\pm 2\pi/3$, or have an axis passing though the midpoints of opposite edges, the angle of rotation being π. There are 8 rotations of the first type and 3 rotations of the second type, consistent with A_4 having order 12.

We now consider the symmetries of **T** which are not rotations. For each edge of **T**, there is a plane of symmetry passing through it, and hence a pure reflection. There are therefore 6 such pure reflections. This leaves us searching for 6 more elements. These are in fact rotated reflections, where the axis for the rotation is a line passing though the midpoints of opposite edges, but the angle of rotation is on this occasion $\pm\pi/2$. Note that neither the rotation about this axis through an angle $\pm\pi/2$, nor the pure reflection in the orthogonal plane, represents a symmetry of **T**, but the composite does.

1.4   Curves and their lengths

Crucial to the study of all the geometries in this course will be the curves lying on them. We consider first the case of a general metric space (X, d), and then we consider the specific case of curves in $\mathbf{R}^n$.

Definition 1.8   A *curve* (or *path*) Γ in a metric space (X, d) is a continuous function $\Gamma : [a, b] \to X$, for some real closed interval $[a, b]$; by an obvious linear reparametrization, we may assume if we wish that $\Gamma : [0, 1] \to X$. A metric space is called *path connected* if any two points of X may be joined by a continuous path.

This is closely related to the concept of a metric (or topological) space X being *connected*; that is, when there is no decomposition of X into the union of two disjoint non-empty open subsets. Equivalently, this is saying that there is no continous function from X onto the two element set $\{0, 1\}$. If there is such a function f, then

$X = f^{-1}(0) \cup f^{-1}(1)$ and X is not connected (it is *disconnected*); conversely, if $X = U_0 \cup U_1$, with U_0, U_1 disjoint non-empty open subsets, then we can define a continuous function f from X onto $\{0, 1\}$, by stipulating that it takes value 0 on U_0 and 1 on U_1. From the definitions, it is easily checked that both connectedness and path connectedness are topological properties, in that they are invariant under homeomorphisms.

If X is path connected, then it is connected. If not, there would be a surjective continuous function $f : X \to \{0, 1\}$; we can then choose points P, Q at which f takes the value 0, 1 respectively, and let Γ be a path joining P to Q. Then $f \circ \Gamma : [a, b] \to \{0, 1\}$ is a surjective continuous function, contradicting the Intermediate Value theorem. All the metric spaces we wish to consider in this course will however have the further property of being *locally path connected*, that is each point of X has a path connected open neighbourhood; for such spaces, it is easy to see conversely that connectedness implies path connectedness (Exercise 1.7), and so the two concepts coincide (although this is not true in general). In particular, the two concepts coincide for open subsets of $\mathbf{R}^n$.

Definition 1.9   For a curve $\Gamma : [a, b] \to X$ on a metric space (X, d), we consider dissections

$$\mathcal{D} : a = t_0 < t_1 < \cdots < t_N = b$$

of $[a, b]$, with N arbitrary. We set $P_i = \Gamma(t_i)$ and $s_{\mathcal{D}} := \sum d(P_i, P_{i+1})$.
The *length* l of Γ is defined to be

$$l = \sup_{\mathcal{D}} s_{\mathcal{D}},$$

if this is finite. For curves in $\mathbf{R}^n$, this is illustrated below.

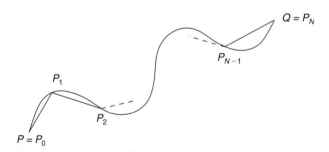

If $\mathcal{D}'$ is a *refinement* of $\mathcal{D}$ (i.e. with extra dissection points), the triangle inequality implies $s_{\mathcal{D}} \le s_{\mathcal{D}'}$. Moreover, given dissections $\mathcal{D}_1$ and $\mathcal{D}_2$, we can find a common refinement $\mathcal{D}_1 \cup \mathcal{D}_2$, by taking the union of the dissection points. Therefore, we may also define the length as $l = \lim_{\mathrm{mesh}(\mathcal{D}) \to 0} s_{\mathcal{D}}$, where by definition $\mathrm{mesh}(\mathcal{D}) = \max_i (t_i - t_{i-1})$. Note that l is the smallest number such that $l \ge s_{\mathcal{D}}$ for all $\mathcal{D}$. By taking the dissection just consisting of a and b, we see that $l \ge d(\Gamma(a), \Gamma(b))$. In the

Euclidean case, any curve joining the two end-points which achieves this minimum length is a straight line segment (Exercise 1.8).

There do exist curves $\Gamma : [a, b] \to \mathbf{R}^2$ (where $[a, b]$ is a finite closed real interval) which fail to have finite length (see for instance Exercise 1.9), but by Proposition 1.10 below this is not the case for sufficiently nice curves. If X denotes a path connected open subset of $\mathbf{R}^n$, it is the case that any two points may be connected by a curve of finite length. This property however fails for example for $\mathbf{R}^2$ with the British Rail metric: this space is certainly path connected, but it is easily checked that any non-constant curve has infinite length.

A metric space (X, d) is called a *length space* if for any two points P, Q of X,

$$d(P, Q) = \inf \{ \text{length}(\Gamma) \ : \ \Gamma \text{ a curve joining } P \text{ to } Q \},$$

and the metric is sometimes called an *intrinsic* metric. In fact, if we start from a metric space (X, d_0) satisfying the property that any two points may be joined by a curve of finite length, then we can define a metric d on X via the above recipe, *defining* $d(P, Q)$ to be the infimum of lengths of curves joining the two points; it is easy to see that this is a metric, and (X, d) is then a length space by Exercise 1.17.

Example   If X denotes a path connected open subset of $\mathbf{R}^2$, and d_0 denotes the Euclidean metric, we obtain an induced intrinsic metric d, where $d(P, Q)$ is the infimum of the lengths of curves in X joining P to Q. Easy examples show that, in general, this is not the Euclidean metric.

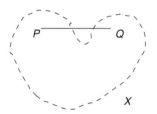

Moreover, the distance $d(P, Q)$ will not in general be achievable as the length of a curve joining P to Q. If for instance $X = \mathbf{R}^2 \setminus \{(0, 0)\}$, then the intrinsic metric d is just the Euclidean metric d_0, but for $P = (-1, 0)$ and $Q = (1, 0)$, there is no curve of length $d(P, Q) = 2$ joining P to Q.

The geometries we study in this course will have underlying metric spaces which are length spaces. Moreover, for most of the important geometries, the space will have the property that the distance between any two points is achieved as the length of some curve joining them; a length space with this property is called a *geodesic space*. This curve of minimum length is often called a *geodesic*, although the definition we give in Chapter 7 will be slightly different (albeit closely related). It might be observed that the London Underground metric (as defined on the appropriate quotient of $\mathbf{R}^2$) determines a geodesic space, in that between any two points there will be a (possibly non-unique) route of minimum length.

Having talked in the abstract about curves on metric spaces, let us now consider the important case of curves in $\mathbf{R}^3$. In the geometries described in this course, we shall usually wish to impose a stronger condition on a curve Γ than just that of continuity; from Chapter 4 onwards, the property of being *piecewise continuously differentiable* will nearly always be the minimum we assume. Given such a curve in $\mathbf{R}^3$, by definition it may be subdivided into a finite number of continuously differentiable parts; to find the length of the curve, we need only find the lengths of these parts. We reduce therefore to the case when Γ is continuously differentiable.

Proposition 1.10   *If $\Gamma : [a,b] \to \mathbf{R}^3$ is continuously differentiable, then*

$$\text{length } \Gamma = \int_a^b \|\Gamma'(t)\| \, dt,$$

where the integrand is the Euclidean norm of the vector $\Gamma'(t) \in \mathbf{R}^3$.

Proof   We write $\Gamma(t) = (f_1(t), f_2(t), f_3(t))$. Thus given $s \neq t \in [a,b]$, the Mean Value theorem implies

$$\frac{\Gamma(t) - \Gamma(s)}{t - s} = (f_1'(\xi_1), f_2'(\xi_2), f_3'(\xi_3))$$

for some $\xi_i \in (s,t)$. Since the f_i' are continuous on $[a,b]$, they are uniformly continuous in the sense of Lemma 1.13: so for any $\varepsilon > 0$, there exists $\delta > 0$ such that, for $1 \leq i \leq 3$,

$$|t - s| < \delta \quad \Longrightarrow \quad |f_i'(\xi_i) - f_i'(\xi)| < \frac{\varepsilon}{3} \qquad \text{for all } \xi \in (s,t).$$

Therefore, if $|t - s| < \delta$, then

$$\|\Gamma(t) - \Gamma(s) - (t - s)\Gamma'(\xi)\| < \varepsilon(t - s) \qquad \text{for all } \xi \in (s,t).$$

Now take a dissection

$$\mathcal{D} : a = t_0 < t_1 < \cdots < t_N = b$$

of $[a,b]$, with $\text{mesh}(\mathcal{D}) < \delta$. The Euclidean distance $d(\Gamma(t_{i-1}), \Gamma(t_i))$ equals $\|\Gamma(t_i) - \Gamma(t_{i-1})\|$. The triangle inequality implies

$$\sum (t_i - t_{i-1})\|\Gamma'(t_{i-1})\| - \varepsilon(b - a) < s_{\mathcal{D}}$$

$$< \sum (t_i - t_{i-1})\|\Gamma'(t_{i-1})\| + \varepsilon(b - a).$$

As $\|\Gamma'(t)\|$ is continuous, it is Riemann integrable and

$$\sum (t_i - t_{i-1})\|\Gamma'(t_{i-1})\| \rightarrow \int_a^b \|\Gamma'(t)\| \, dt$$

as mesh$(\mathcal{D}) \rightarrow 0$. Therefore

$$\text{length } \Gamma := \lim_{\text{mesh}(\mathcal{D})\rightarrow 0} s_{\mathcal{D}}$$

$$= \int_a^b \|\Gamma'(t)\| \, dt. \qquad \square$$

For (piecewise) continuously differentiable curves $\Gamma : [a,b] \rightarrow \mathbf{R}^n$, this proof extends immediately to show that we may obtain the length of such curves by integrating $\|\Gamma'\|$.

1.5 Completeness and compactness

We should mention two further well-known conditions on metric spaces, one defined metrically and the other topologically, namely completeness and compactness. Some readers may be familiar with these concepts, and should therefore just omit this section. The account given here will only be a brief sketch of what is standard theory; the reader should refer to a suitable book, such as [13], for more details.

Definition 1.11   A sequence $x_1, x_2, \ldots$ of points in a metric space (X, d) is called a *Cauchy sequence* if, for any $\varepsilon > 0$, there exists an integer N such that if $m, n \geq N$ then $d(x_m, x_n) < \varepsilon$. We say that the space is *complete* if every Cauchy sequence (x_n) has a limit in X, that is a point $x \in X$ such that $d(x_n, x) \rightarrow 0$ as $n \rightarrow \infty$. Such limits are clearly unique.

It is a well-known fact that real Cauchy sequences converge, and so the real line (with its standard metric) is complete. Applying this to the coordinates of points in $\mathbf{R}^n$, we deduce easily that the Euclidean space $\mathbf{R}^n$ is also complete. A subset X of $\mathbf{R}^n$ will be complete if and only if it is closed, since an equivalent condition for a subset to be closed is that it contains all its limit points.

Thus, the open unit disc D in $\mathbf{R}^2$ is not complete. We saw however that D is homeomorphic to $\mathbf{R}^2$, which is complete. Therefore, completeness is not a topological property, but depends on the metric. Most of the geometries we study in this course will have this property of completeness.

The other property which will make occasional appearances in later chapters is that of compactness; the definition of compactness is phrased purely in terms of open sets, and so it is a topological property.

Definition 1.12   Given a metric space X (or more generally, for readers who know about them, a topological space), we say that X is *compact* if any cover of X by open subsets has a finite subcover. A *cover* of X by open subsets is a collection of open subsets $\{U_i\}_{i \in I}$, for an arbitrary indexing set I, whose union is all of X. Compactness says that there will always be a finite subcollection of these sets whose union is all of X.

A standard result for *metric spaces* is that compactness is equivalent to another condition, called sequential compactness ([13], Chapter 7). A metric space (X, d) is called *sequentially compact* if every sequence in X has a convergent subsequence. A well-known basic result in elementary analysis says that a finite closed interval $[a, b]$ of the real line is compact. With our topological definition of compactness, this is the Heine–Borel theorem. With the characterization of compactness via sequential compactness, this is the Bolzano–Weierstrass theorem. This result may be generalized in a straightforward manner to $\mathbf{R}^n$ (for instance, by applying the Bolzano–Weierstrass theorem to the components of vectors), to deduce that any closed box $[a_1, b_1] \times \cdots \times [a_n, b_n]$ in $\mathbf{R}^n$ is compact.

Using the interpretation of compactness in terms of sequential compactness, we observe that a compact metric space X is necessarily complete. Indeed, if X is not complete, we could take a Cauchy sequence (x_n) with no limit point in X. If there were a convergent subsequence, then the Cauchy condition would show that the whole sequence was convergent, contradicting our initial choice. The Euclidean plane $\mathbf{R}^2$, and when we come to it later, the hyperbolic plane, are examples of metric spaces which are complete but not compact.

A very useful property of compact metric spaces is the following fact concerning continuous functions.

Lemma 1.13   *A continuous function $f : X \to \mathbf{R}$ on a compact metric space (X, d) is uniformly continuous, i.e. given $\varepsilon > 0$, there exists $\delta > 0$ such that if $d(x, y) < \delta$, then $|f(x) - f(y)| < \varepsilon$.*

Proof   For each $x \in X$, there exists $\delta(x) > 0$ such that if $d(y, x) < 2\delta(x)$, then $|f(y) - f(x)| < \varepsilon/2$. Since X is covered by the open balls $B(x, \delta(x))$, by compactness it is covered by finitely many of them, say $B(x_i, \delta(x_i))$, with $i = 1, \ldots, n$. Let $\delta = \min_i \{\delta(x_i)\}$. Suppose now that $d(x, y) < \delta$; there exists i such that $x \in B(x_i, \delta(x_i))$, and so $y \in B(x_i, 2\delta(x_i))$. Therefore $|f(x) - f(x_i)| < \varepsilon/2$ and $|f(y) - f(x_i)| < \varepsilon/2$, which implies that $|f(x) - f(y)| < \varepsilon$. $\square$

Finally in this section, we prove a further two elementary results on compactness.

Lemma 1.14   *If Y is a closed subset of a compact metric (or topological) space X, then Y is compact.*

Proof   The open subsets of Y are of the form $U \cap Y$, for U an open subset of X. When X is a metric space, this is just because the open subsets are characterized by being the union of open balls, and the restriction of an open ball in X centred on $y \in Y$ is an open ball in Y; for topological spaces, this is true by definition.

Suppose now we have an open cover $\{V_i\}_{i \in I}$ of Y, and write each $V_i = U_i \cap Y$ for an appropriate open subset U_i of X. The union of these open sets U_i therefore contains Y, and hence these open sets together with the open set $X \setminus Y$ cover X. The compactness of X implies that there exists a finite subcover, which in turn implies that for some finite subcollection of the U_i, the union contains Y. This then says that some finite subcollection of the V_i cover Y. □

Combining this with previous results, we deduce that any closed and bounded subset X of $\mathbf{R}^n$ is compact, since X is a closed subset of some closed box in $\mathbf{R}^n$. It is straightforward to check that the converse is true; any compact subset of $\mathbf{R}^n$ is closed and bounded (Exercise 1.10). Thus, for instance, the unit sphere $S^n \subset \mathbf{R}^{n+1}$ is seen to be compact.

Lemma 1.15   *If $f : X \to Y$ is a continuous surjective map of metric (or topological) spaces, with X compact, then so too is Y.*

Proof   This follows directly from the definition of compactness. Suppose we have an open cover $\{U_i\}_{i\in I}$ of Y. Then $\{f^{-1}U_i\}_{i\in I}$ is an open cover of X, and so by assumption has a finite subcover. The surjectivity condition then implies that the corresponding finite subcollection of the U_i cover Y. □

With the aid of this last lemma, Exercise 1.10 yields the well-known result that a continuous real-valued function on a compact metric space is bounded and attains its bounds.

When, in later chapters, we study the torus, it may be seen to be compact in two different ways: either because it may be realized as a closed bounded subset of $\mathbf{R}^3$, or because there is a continuous surjective map to it from a closed square in $\mathbf{R}^2$.

1.6   Polygons in the Euclidean plane

A key concept in later chapters will be that of geodesic polygons. In this section, we shall characterize Euclidean polygons in $\mathbf{R}^2$ as the 'inside' of a simple closed polygonal curve, although the results we prove will be more generally applicable.

Definition 1.16   For X a metric space, a curve $\gamma : [a, b] \to X$ is called *closed* if $\gamma(a) = \gamma(b)$. It is called *simple* if, for $t_1 < t_2$, we have $\gamma(t_1) \neq \gamma(t_2)$, with the possible exception of $t_1 = a$ and $t_2 = b$, when the curve is closed.

The famous example here is that of simple closed curves in $\mathbf{R}^2$, for which the Jordan Curve theorem states that the complement of the curve in $\mathbf{R}^2$ consists of precisely two path connected components, a bounded component (called the *inside* of γ) and an unbounded component (called the *outside*). In general, when the continuous curve γ may be highly complicated (it may for instance look locally like the curve in Exercise 1.9), this is a difficult result. We shall prove it below for the simple case when γ is *polygonal*, meaning that it consists of a finite number of straight line segments, but the proof given will turn out to be applicable to other cases we need. The proof comes in two parts: firstly, we show that the complement has at most two connected components, and then we show that there are precisely two components.

Proposition 1.17   *Let $\gamma : [a, b] \to \mathbf{R}^2$ be a simple closed polygonal curve, with $C \subset \mathbf{R}^2$ denoting the image $\gamma([a, b])$. Then $\mathbf{R}^2 \setminus C$ has at most two path connected components.*

Proof   For each $P \in C$, we can find an open ball $B = B(P, \varepsilon)$ such that $C \cap B$ consists of two radial linear segments (often a diameter). The set C is compact by Lemma 1.15, and is contained in the union of such open balls. Arguing as in the proof of Lemma 1.14, we deduce that C is contained in the union U of some finite subcollection of these balls, say $U = B_1 \cup \cdots \cup B_N$.

Our assumptions imply that $U \setminus C$ consists of two path connected components U_1 and U_2; if one travels along the curve γ, then one of these components will always be to the left, and the other always to the right. The fact that U is a finite union of the balls ensures that, given any two points P, Q of U_1, there is a path in U_1 joining the points, as illustrated below, and that a similar statement holds for U_2.

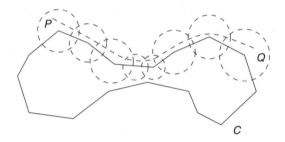

Suppose now that we have arbitrary points P, Q of $\mathbf{R}^2 \setminus C$; for each point, we take a path (for instance a straight line path) joining the point to a point of C. In both cases, just before we reach C for the first time, we will be in one of the open sets U_1 or U_2. If we are in the *same* U_i for the paths starting from both P and Q, then the path connectedness of U_i ensures that there is a path from P to Q in $\mathbf{R}^2 \setminus C$. Thus $\mathbf{R}^2 \setminus C$ has at most two path connected components. $\square$

Remark 1.18   The above proof is far more widely applicable than just to the case when the simple closed curve is polygonal. It clearly extends for instance to the case when γ is made up of circular arcs and line segments. More generally still, it also applies under the assumption that every point $P \in C$ has an open neighbourhood V which is *homeomorphic* to an open disc B in $\mathbf{R}^2$ such that the curve in B corresponding to $C \cap V$ consists of two radial linear segments (including the case of a diameter). This last observation will be used in the proof of Theorem 8.15.

The fact that for simple closed polygonal curves in $\mathbf{R}^2$ (and similarly nice curves) there is more than one component, follows most easily from an argument involving *winding numbers*. Winding numbers will be used, in this book, to identify, in a rigorous way, the *inside* of suitably well-behaved simple closed curves. Here is not the place to give a full exposition on winding numbers. The reader who has taken a course on Complex Analysis should be familiar with them; for other readers, I describe briefly their salient properties. For full details, the reader is referred to Section 7.2 of [1].

Given a set $A \subset \mathbf{C}^* = \mathbf{C} \setminus \{0\}$, a *continuous branch of the argument* on A is a continuous function $h : A \to \mathbf{R}$ such that $h(z)$ is an argument of z for all $z \in A$. If for instance $A \subset \mathbf{C} \setminus \mathbf{R}_{\geq 0} e^{i\alpha}$ for some $\alpha \in \mathbf{R}$, where $\mathbf{R}_{\geq 0} = \{r \in \mathbf{R} : r \geq 0\}$, then clearly such an h may be chosen on A with values in the range $(\alpha, \alpha + 2\pi)$. On the other hand, one cannot choose a continuous branch of the argument on the whole of $\mathbf{C}^*$, and this is the basic reason for the existence of winding numbers. Note that a continuous branch of the argument exists on A if and only if a *continuous branch of the logarithm* exists, i.e. a continuous function $g : A \to \mathbf{R}$ such that $\exp g(z) = z$ for all $z \in A$, since the functions h and g may be obtained from each other (modulo perhaps an integral multiple of $2\pi i$) via the relation $g(z) = \log |z| + ih(z)$.

Suppose now that $\gamma : [a, b] \to \mathbf{C}^*$ is any curve; a *continuous branch of the argument* for γ is a continuous function $\theta : [a, b] \to \mathbf{R}$ such that $\theta(t)$ is an argument for $\gamma(t)$ for all $t \in [a, b]$. Note that two different continuous branches of the argument, say θ_1 and θ_2, have the property that $(\theta_1 - \theta_2)/2\pi$ is a continuous integer valued function on $[a, b]$, and hence constant by the Intermediate Value theorem. Thus two different continuous branches of the argument for γ just differ by an integral multiple of 2π. Unlike continuous branches of the argument for subsets, continuous branches of the argument for curves in $\mathbf{C}^*$ always exist. Using the continuity of the curve, one sees easily that they exist *locally* on $[a, b]$, and one can then use the compactness of $[a, b]$ to achieve a continuous function on the whole of $[a, b]$ (see [1], Theorem 7.2.1).

If now $\gamma : [a, b] \to \mathbf{C}^*$ is a *closed* curve, we define the *winding number* or *index* of γ about the origin, denoted $n(\gamma, 0)$, by choosing any continuous branch of the argument θ for γ, and letting $n(\gamma, 0)$ be the well-defined integer given by

$$n(\gamma, 0) = (\theta(b) - \theta(a))/2\pi.$$

More generally, given any closed curve $\gamma : [a, b] \to \mathbf{C} = \mathbf{R}^2$ and a point w not on the curve, we define the winding number of γ about w to be the integer $n(\gamma, w) := n(\gamma - w, 0)$, where $\gamma - w$ is the curve whose value at $t \in [a, b]$ is $\gamma(t) - w$. Intuitively, this integer $n(\gamma, w)$ describes how many times (and in which direction) the curve γ 'winds round w'. If, for instance, γ denotes the boundary of a triangle traversed in an anti-clockwise direction and w is a point in the interior of the triangle, one checks easily that $n(\gamma, w) = 1$. On the other hand, if we take γ in the clockwise direction, then $n(\gamma, w) = -1$.

Some elementary properties (see [1], Section 7.2) of the winding number of a closed curve γ include the following.

- If we reparametrize γ or choose a different starting point on the curve, the winding number is unchanged. However, if $-\gamma$ denotes the curve γ travelled in the opposite direction, i.e. $(-\gamma)(t) = \gamma(b - (b - a)t)$, then for any w not on the curve,

$$n((-\gamma), w) = -n(\gamma, w).$$

For the constant curve γ, we have $n(\gamma, w) = 0$.

- If the curve $\gamma - w$ is contained in a subset $A \subset \mathbf{C}^*$ on which a continuous branch of the argument can be defined, then $n(\gamma, w) = 0$. From this, it follows easily that if γ is contained in a closed ball $\bar{B}$, then $n(\gamma, w) = 0$ for all $w \notin \bar{B}$.
- As a function of w, the winding number $n(\gamma, w)$ is constant on each path connected component of the complement of $C := \gamma([a, b])$.
- If $\gamma_1, \gamma_2 : [0, 1] \to \mathbf{C}$ are two closed curves with $\gamma_1(0) = \gamma_1(1) = \gamma_2(0) = \gamma_2(1)$, we can form the *concatenation* $\gamma = \gamma_1 * \gamma_2 : [0, 2] \to \mathbf{C}$, defined by

$$\gamma(t) = \begin{cases} \gamma_1(t) & \text{for } 0 \le t \le 1, \\ \gamma_2(t-1) & \text{for } 1 \le t \le 2. \end{cases}$$

Then for w not in the image of $\gamma_1 * \gamma_2$, we have

$$n(\gamma_1 * \gamma_2, w) = n(\gamma_1, w) + n(\gamma_2, w).$$

Proposition 1.19   *If $\gamma : [a, b] \to \mathbf{C} = \mathbf{R}^2$ is a polygonal simple closed curve, there exist points w, not in the image C of γ, for which $n(\gamma, w) = \pm 1$. Hence, in view of the previous result, there exist precisely two path connected components of the complement of C.*

Proof   Consider the continuous function $\|\gamma(t)\|$ for $t \in [a, b]$. This is a continuous function on a closed interval, and so attains its bounds by Exercise 1.10. There exists therefore a point $P_2 \in C$ with $d(0, P_2)$ maximum; from this it is clear that P_2 is a vertex of the polygonal curve. If there is more than one point of the curve at maximum distance d from the origin, we just choose one of them to be P_2 and shift the origin a small distance ε to $0'$ as shown, to ensure the uniqueness of P_2. All the points of C are within the closed disc of radius d centred on 0, and so all the points except P_2 are in the open disc of radius $d + \varepsilon$ centred at $0'$.

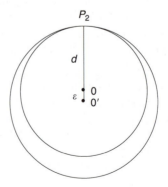

Set P_1, P_3 to denote the vertices immediately before (respectively, after) P_2, and let l denote the line segment from P_1 to P_3, suitably parametrized.

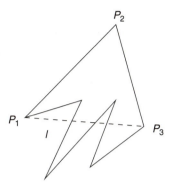

We let γ_1 denote the polygonal closed curve (no longer simple, in general) obtained from γ by missing out the vertex P_2 and going straight from P_1 to P_3 via l. Clearly, γ_1 is contained in some closed ball $\bar{B}(0, \delta)$ with $\delta < d(0, P_2)$, and hence $n(\gamma_1, w) = 0$ for all points w sufficiently close to P_2. We let γ_2 denote the triangular path $P_3P_1P_2P_3$, say with the parametrization of the segment from P_3 to P_1 being the opposite to that chosen for the segment P_1P_3 in γ_1. Elementary properties of the winding number give, for all w not in the image of $\gamma_1 * \gamma_2$, that $n(\gamma_1 * \gamma_2, w) = n(\gamma, w)$ (since the two contributions from the segment l are in opposite directions, and therefore cancel). Moreover, it follows immediately from the definition of the winding number that, for any w in the interior of the triangle $P_1P_2P_3$, the winding number $n(\gamma_2, w) = \pm 1$, the sign being plus if γ_2 goes round the perimeter in an anti-clockwise direction. Putting these facts together, for w in the interior of the triangle *and* sufficiently close to P_2, we have

$$n(\gamma, w) = n(\gamma_1 * \gamma_2, w) = n(\gamma_1, w) + n(\gamma_2, w) = \pm 1.$$

Since γ is contained in a closed ball, we have that $n(\gamma, w) = 0$ for w outside that ball, and so the result follows from the previous result and the third stated property for winding numbers. $\qquad\square$

Remark 1.20 The above proof is again more widely applicable than just to the case of simple closed polygonal curves. Suppose, for instance, that γ is made up of circular arcs and line segments. As before, we choose a point $P_2 \in C$ with $d(0, P_2)$ maximum, which by the same argument as above may be assumed unique. If P_2 is not a vertex, it must be an interior point of a circular arc between vertices P_1 and P_3; in the case where it is a vertex, we let P_1, P_3 denote the vertices immediately before (respectively, after) P_2. As before l will denote the line segment from P_1 to P_3. Whether P_2 is a vertex or not, the above argument still works, where γ_1 denotes the curve obtained from γ, but going straight from P_1 to P_3 along l, and γ_2 is the path going from P_1 to P_3 via P_2 (along the curve γ) and returning to P_1 along l.

Definition 1.21 For a simple closed polygonal curve with image $C \subset \mathbf{R}^2$, we know that C is compact, and hence bounded by Exercise 1.10. Thus C is contained in some closed ball $\bar{B}$. Since any two points in the complement of $\bar{B}$ may be joined by a path, one of the two components of $\mathbf{R}^2 \setminus C$ contains the complement of $\bar{B}$, and hence is *unbounded*, whilst the other component of $\mathbf{R}^2 \setminus C$ is contained in $\bar{B}$, and hence is *bounded*. The closure of the bounded component, which consists of the bounded component together with C, will be called a *closed polygon* in $\mathbf{R}^2$ or a *Euclidean polygon*. Since a Euclidean polygon is closed and bounded in $\mathbf{R}^2$, it too is compact.

Exercises

1.1   If ABC is a triangle in $\mathbf{R}^2$, show that the three perpendicular bisectors of the sides meet at a point O, which is the centre of a circle passing though A, B and C.

1.2   If f is an isometry of $\mathbf{R}^n$ to itself which fixes all points on some affine hyperplane H, show that f is either the identity or the reflection R_H.

1.3   Let l, l' be two distinct lines in $\mathbf{R}^2$, meeting at a point P with an angle α. Show that the composite of the corresponding reflections $R_l R_{l'}$ is a rotation about P through an angle 2α. If l, l' are parallel lines, show that the composite is a translation. Give an example of an isometry of $\mathbf{R}^2$ which cannot be expressed as the composite of *less than* three reflections.

1.4   Let $R(P, \theta)$ denote the clockwise rotation of $\mathbf{R}^2$ through an angle θ about a point P. If A, B, C are the vertices, labelled clockwise, of a triangle in $\mathbf{R}^2$, prove that the composite $R(A, \theta)R(B, \phi)R(C, \psi)$ is the identity if and only if $\theta = 2\alpha$, $\phi = 2\beta$ and $\psi = 2\gamma$, where α, β, γ denote the angles at, respectively, the vertices A, B, C of the triangle ABC.

1.5   Let G be a finite subgroup of $\mathrm{Isom}(\mathbf{R}^m)$. By considering the barycentre (i.e. average) of the orbit of the origin under G, or otherwise, show that G fixes some point of $\mathbf{R}^m$. If G is a finite subgroup of $\mathrm{Isom}(\mathbf{R}^2)$, show that it is either cyclic or *dihedral* (that is, $D_4 = C_2 \times C_2$, or, for $n \geq 3$, the full symmetry group D_{2n} of a regular n-gon).

1.6   Show that the interior of a Euclidean triangle in $\mathbf{R}^2$ is homeomorphic to the open unit disc.

1.7   Let (X, d) denote a metric space. Suppose that every point of X has an open neighbourhood which is path connected. If X is connected, prove that it is in fact path connected. Deduce that a connected open subset of $\mathbf{R}^n$ is always path connected. In this latter case, show that any two points may be joined by a polygonal curve.

1.8   Prove that a continuous curve of shortest length between two points in Euclidean space is a straight line segment, parametrized monotonically.

1.9   Show that the plane curve $\gamma : [0, 1] \to \mathbf{R}^2$, defined by $\gamma(t) = (t, t \sin(1/t))$ for $t > 0$ and $\gamma(0) = (0, 0)$, does not have finite length.

1.10 Using sequential compactness, show that any compact subset of $\mathbf{R}^n$ is both closed and bounded. Deduce that a continuous real-valued function on a compact metric space is bounded and attains its bounds.

1.11 Suppose that $f : [0, 1] \to \mathbf{R}$ is a continuous map; show that its image is a closed interval of $\mathbf{R}$. If, furthermore, f is injective, prove that f is a homeomorphism onto its image.

1.12   Let R be a plane polygon. By considering a point of R at maximum distance from the origin, show that there exists at least one vertex of R at which the interior angle is $< \pi$.

1.13   Suppose z_1, z_2 are distinct points of $\mathbf{C}^*$, and that $\Gamma : [0, 1] \to \mathbf{C}^*$ is a (continuous) curve with $\Gamma(0) = z_1$ and $\Gamma(1) = z_2$. Suppose furthermore that, for any $0 \le t \le 1$, the ray $\arg(z) = \arg(\Gamma(t))$ meets Γ only at $\Gamma(t)$. Let γ denote the simple closed curve obtained by concatenating the two radial segments $[0, z_1]$ and $[z_2, 0]$ with Γ. Prove that the complement of this curve in $\mathbf{C}$ consists of precisely two connected components, one bounded and one unbounded, where the closure of the bounded component is the union of the radial segments from 0 to $\Gamma(t)$, for $0 \le t \le 1$.

1.14   Show that the bounded component of the complement in $\mathbf{C}$ of the simple closed curve γ from the previous exercise is homeomorphic to the interior of a Euclidean triangle, and hence, by Exercise 1.6, it is homeomorphic to an open disc in $\mathbf{R}^2$. [This is a general fact about the bounded component of the complement of any simple closed curve in $\mathbf{R}^2$.]

1.15   For a cube centred on the origin in $\mathbf{R}^3$, show that the rotation group is isomorphic to S_4, considered as the permutation group of the four long diagonals. Prove that the full symmetry group is isomorphic to $C_2 \times S_4$, where C_2 is the cyclic group of order 2. How many of the isometries is this group are rotated reflections (and not pure reflections)? Describe these rotated reflections geometrically, by identifying the axes of rotation and the angles of rotation.

1.16   Given F a closed subset of a metric space (X, d), show that the real-valued function $d(x, F) := \inf\{d(x, y) : y \in F\}$ is continuous, and strictly positive on the complement of F. If $K \subset X$ is a compact subset of X, disjoint from F, deduce from Exercise 1.10 that the distance

$$d(K, F) := \inf\{d(x, y) : x \in K, y \in F\}$$

is strictly positive. [We call $d(x, F)$ the *distance* of x from F, and $d(K, F)$ the *distance* between the two subsets.]

1.17   Suppose (X, d_0) is a metric space in which any two points may be joined by a curve of finite length, and let d denote the associated metric, defined as in Section 1.4 via lengths of curves. For any curve $\gamma : [a, b] \to X$, we denote by $l_{d_0}(\gamma)$, respectively $l_d(\gamma)$, the lengths of γ as defined with respect to the two metrics.

(a)   Show that $d_0(P, Q) \le d(P, Q)$ for all $P, Q \in X$; deduce that $l_{d_0}(\gamma) \le l_d(\gamma)$.

(b)   For any dissection $\mathcal{D} : a = t_0 < t_1 < \cdots < t_N = b$ of $[a, b]$, show that

$$d(\gamma(t_{i-1}), \gamma(t_i)) \le l_{d_0}(\gamma|_{[t_{i-1}, t_i]})$$

for $1 \le i \le N$.

(c)   By summing the inequalities in (b), deduce that $l_d(\gamma) \le l_{d_0}(\gamma)$, and hence by (a) that the two lengths are equal.

Deduce that d is an intrinsic metric.

2 Spherical geometry

Introduction

We now study our first non-Euclidean two-dimensional geometry, namely the geometry that arises on the surface of a sphere. Intuitively, this should be no more difficult for us to visualize than the Euclidean plane, since to a first approximation we do live on the surface of a sphere, namely the Earth, and we are used to making journeys in this geometry, which correspond to curves on the sphere. We shall normalize this geometry so that the sphere has unit radius. In this chapter, we let $S = S^2$ denote the unit sphere in $\mathbf{R}^3$ with centre $O = \mathbf{0}$, and we use the two notations interchangeably.

A *great circle* on S is the intersection of S with a plane through the origin. We shall refer to great circles as (spherical) *lines* on S. Through any two non-antipodal points P, Q on S, there exists precisely one line (namely, we intersect S with the plane determined by OPQ).

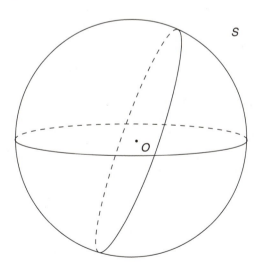

Definition 2.1   The *distance* $d(P, Q)$ between P and Q on S is defined to be the length of the shorter of the two segments PQ along the great circle (where this is π if

P and Q are antipodal). In this chapter, we shall always use d to denote this distance function on the sphere.

Note that $d(P, Q)$ is just the angle between $\mathbf{P} = \overrightarrow{OP}$ and $\mathbf{Q} = \overrightarrow{OQ}$, and hence is just $\cos^{-1}(\mathbf{P}, \mathbf{Q})$, where $(\mathbf{P}, \mathbf{Q}) = \mathbf{P} \cdot \mathbf{Q}$ is the Euclidean inner-product on $\mathbf{R}^3$. For reasons which will become clear later, the spherical lines are also sometimes called the geodesics or geodesic lines on S^2.

2.2 Spherical triangles

Definition 2.2   A spherical triangle ABC on S is defined by its vertices $A, B, C \in S$, and sides AB, BC and AC, where these are spherical line segments on S of length $< \pi$.

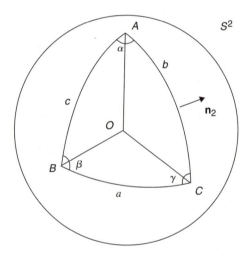

The triangle ABC is the region of the sphere with area $< 2\pi$ enclosed by these sides — our assumption on the length of the sides is equivalent to the assumption that the triangle is contained in some (open) hemisphere (Exercise 2.3).

Setting $\mathbf{A} = \overrightarrow{OA}$, $\mathbf{B} = \overrightarrow{OB}$ and $\mathbf{C} = \overrightarrow{OC}$, the length of the side AB is given by $c = \cos^{-1}(\mathbf{A} \cdot \mathbf{B})$, with similar formulae for the lengths a, b of the sides BC and CA, respectively. Denoting the cross-product of vectors in $\mathbf{R}^2$ by $\times$, we now set

$$\mathbf{n}_1 = \mathbf{C} \times \mathbf{B} / \sin a,$$

$$\mathbf{n}_2 = \mathbf{A} \times \mathbf{C} / \sin b,$$

$$\mathbf{n}_3 = \mathbf{B} \times \mathbf{A} / \sin c,$$

the unit normals to the planes OBC, OAC, OBA (these normals will be pointing out of the solid $OABC$, provided we have labelled our vertices anticlockwise). The angles of the spherical triangle are defined to be the angles between the defining planes for the sides.

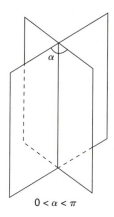

$$0 < \alpha < \pi$$

Noting that the angle between $\mathbf{n}_2$ and $\mathbf{n}_3$ is $\pi + \alpha$, i.e. the non-reflex angle is $\pi - \alpha$, we have that $\mathbf{n}_2 \cdot \mathbf{n}_3 = -\cos\alpha$. Similarly, $\mathbf{n}_3 \cdot \mathbf{n}_1 = -\cos\beta$ and $\mathbf{n}_1 \cdot \mathbf{n}_2 = -\cos\gamma$.

Theorem 2.3 (Spherical cosine formula)

$$\sin a \, \sin b \, \cos\gamma = \cos c - \cos a \, \cos b.$$

Proof We use the vector identity

$$(\mathbf{C} \times \mathbf{B}) \cdot (\mathbf{A} \times \mathbf{C}) = (\mathbf{A} \cdot \mathbf{C})(\mathbf{B} \cdot \mathbf{C}) - (\mathbf{C} \cdot \mathbf{C})(\mathbf{B} \cdot \mathbf{A}).$$

In our case, $|\mathbf{C}| = 1$ and so the right-hand side is just $(\mathbf{A} \cdot \mathbf{C})(\mathbf{B} \cdot \mathbf{C}) - (\mathbf{B} \cdot \mathbf{A})$. Therefore

$$
\begin{aligned}
- \sin a \, \sin b \, \cos\gamma &= \sin a \, \sin b \, \mathbf{n}_1 \cdot \mathbf{n}_2 \\
&= (\mathbf{C} \times \mathbf{B}) \cdot (\mathbf{A} \times \mathbf{C}) \\
&= (\mathbf{A} \cdot \mathbf{C})(\mathbf{B} \cdot \mathbf{C}) - (\mathbf{B} \cdot \mathbf{A}) \\
&= \cos b \, \cos a - \cos c. \qquad \square
\end{aligned}
$$

As for Euclidean triangles, we obtain the spherical Pythagoras theorem as a special case of the cosine formula.

Corollary 2.4 (Spherical Pythagoras theorem)   *When $\gamma = \frac{\pi}{2}$,*

$$\cos c = \cos a \, \cos b. \qquad \square$$

There is also a formula corresponding to the Euclidean sine formula.

Theorem 2.5 (Spherical sine formula)   *With the notation as above,*

$$\frac{\sin a}{\sin\alpha} = \frac{\sin b}{\sin\beta} = \frac{\sin c}{\sin\gamma}.$$

Proof   We use the vector identity

$$(\mathbf{A} \times \mathbf{C}) \times (\mathbf{C} \times \mathbf{B}) = (\mathbf{C} \cdot (\mathbf{B} \times \mathbf{A}))\mathbf{C}.$$

In our case, the left-hand side of this equation is $-(\mathbf{n}_1 \times \mathbf{n}_2) \sin a \sin b$. Clearly $\mathbf{n}_1 \times \mathbf{n}_2$ is a multiple of $\mathbf{C}$, and one verifies easily that $\mathbf{n}_1 \times \mathbf{n}_2 = \mathbf{C} \sin \gamma$. Therefore, equating the multiples of $\mathbf{C}$, we deduce that

$$\mathbf{C} \cdot (\mathbf{A} \times \mathbf{B}) = \sin a \, \sin b \, \sin \gamma.$$

The triple product being invariant under cyclic permutations, we get

$$\sin a \, \sin b \, \sin \gamma = \sin b \, \sin c \, \sin \alpha = \sin c \, \sin a \, \sin \beta.$$

Dividing by $\sin a \, \sin b \, \sin c$, we obtain the result.   □

Remark 2.6

(i)   For a, b, c small, these formulae reduce to the Euclidean versions in the limit. For example, in (2.3) we have

$$ab \cos \gamma = \left(1 - \frac{c^2}{2}\right) - \left(1 - \frac{a^2}{2}\right)\left(1 - \frac{b^2}{2}\right) + O(3),$$

whose second order terms yield the Euclidean cosine formula.

(ii)   We may avoid the use of vector identities by rotating the sphere so that the vertex A say lies at the north pole, thus corresponding to the column vector $(0, 0, 1)^t$. We shall adopt this approach later when proving the hyperbolic versions of these formulae, where the corresponding vector identities are slightly trickier and less familiar.

Assuming that $a, b, c < \pi$, Theorem 2.3 implies that

$$\cos c = \cos a \, \cos b + \sin a \, \sin b \, \cos \gamma.$$

Thus, unless $\gamma = \pi$ (i.e. C lies on the line segment AB and hence $c = a + b$),

$$\cos c > \cos a \, \cos b - \sin a \, \sin b = \cos(a + b),$$

and so $c < a + b$.

Corollary 2.7 (Triangle inequality)   *For $P, Q, R \in S^2$,*

$$d(P, Q) + d(Q, R) \geq d(P, R),$$

with equality if and only if Q is on the line segment PR (of shorter length). In particular, it follows that the distance function d that has been defined is a metric, the spherical metric.

Proof We now only need to worry about the case $d(P, R) = \pi$, i.e. P and R antipodal. In that case, the line PQ on S also passes through R and so

$$d(P, R) = d(P, Q) + d(Q, R).$$

$\square$

In contrast to the Euclidean case, there is a rather useful second cosine formula, which may be used for instance to deduce the lengths of the sides of a spherical triangle from knowledge of its angles.

Proposition 2.8 (Second cosine formula) *With the notation as before,*

$$\sin \alpha \, \sin \beta \, \cos c = \cos \gamma - \cos \alpha \, \cos \beta.$$

Proof With the conventions as before concerning the spherical triangle ABC, we denote by $\mathbf{A}'$, $\mathbf{B}'$ and $\mathbf{C}'$ the unit normals $-\mathbf{n}_1$, $-\mathbf{n}_2$, $-\mathbf{n}_3$ in the directions of, respectively, $\mathbf{B} \times \mathbf{C}$, $\mathbf{C} \times \mathbf{A}$ and $\mathbf{A} \times \mathbf{B}$ — these are the inward pointing normals to the solid $OABC$. The corresponding points A', B' and C' on the sphere form the vertices of a spherical triangle $A'B'C'$, called the *polar triangle* to the original triangle ABC. The angle between $\mathbf{B}'$ and $\mathbf{C}'$ is $\pi - \alpha$, and so the side length is $\pi - \alpha$. The other two side lengths are $\pi - \beta$ and $\pi - \gamma$.

To find the angles of the polar triangle, we observe that the polar of the polar triangle is our original triangle; it is clear for instance that the unit vector in the direction $\mathbf{B}' \times \mathbf{C}'$ must be $\pm \mathbf{A}$, and one then easily verifies that it is $\mathbf{A}$. The original triangle had side lengths a, b and c; thus the angles of the polar triangle have to be $\pi - a, \pi - b$ and $\pi - c$. The second cosine formula is then deduced simply by applying the first cosine formula to the polar triangle.

$\square$

2.3 Curves on the sphere

We now have two natural metrics defined on the sphere, the restriction to S of the Euclidean metric on $\mathbf{R}^3$, and the spherical distance metric defined in the previous section. Given a curve $\Gamma : [a, b] \rightarrow S^2$, we can define a length (using the recipe described in Definition 1.10), taking either of these metrics on S^2 as a starting point.

Proposition 2.9 *Given a curve Γ on S joining points P, Q on S, these two concepts of length coincide.*

Proof Suppose $\Gamma : [a, b] \rightarrow S$ is a curve which has length l, when considered as a curve in $\mathbf{R}^3$ with the Euclidean metric. For any given dissection $\mathcal{D}$ of $[a, b]$, where $a = t_0 < t_1 < \cdots < t_N = b$, we set $P_i = \Gamma(t_i)$ and

$$\tilde{s}_\mathcal{D} := \sum_{i=1}^{N} d(P_{i-1}, P_i) > s_\mathcal{D} = \sum_{i=1}^{N} \|\overrightarrow{P_{i-1}P_i}\|.$$

The length of Γ with respect to the spherical metric is just $l' = \sup_{\mathcal{D}} \tilde{s}_{\mathcal{D}}$; this is clearly $\geq l$, and we show that the opposite inequality is also true, and hence that $l = l'$.

We suppose therefore that $l < l'$ and obtain a contradiction. If this inequality holds, we choose $\varepsilon > 0$ such that $(1 + \varepsilon)l < l'$. Since $\frac{\sin\theta}{\theta} \to 1$ as $\theta \to 0$, we have $2\theta \leq (1 + \varepsilon)2\sin\theta$ for θ sufficiently small.

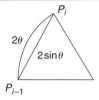

By uniform continuity of Γ, and by taking a sufficiently small mesh, we can therefore choose our dissection $\mathcal{D}$ (for some N sufficiently large) such that

$$d(P_{i-1}, P_i) \leq (1 + \varepsilon)\|\overrightarrow{P_{i-1}P_i}\|,$$

for all $1 \leq i \leq N$. For such a dissection, it follows that

$$\tilde{s}_{\mathcal{D}} \leq (1 + \varepsilon)s_{\mathcal{D}} < (1 + \varepsilon)l.$$

Taking suprema over all dissections, we deduce that $l' \leq (1 + \varepsilon)l < l'$, which is the required contradiction. $\qquad\square$

Proposition 2.10   *Given a curve Γ on S joining points P and Q, we have $l = $ length $\Gamma \geq d(P, Q)$. Moreover, if $l = d(P, Q)$, the image of Γ is the spherical line segment PQ on S.*

Proof   For any curve Γ on S from P to Q, the above discussion shows that its length l is given by $\sup_{\mathcal{D}} \tilde{s}_{\mathcal{D}}$. By taking the dissection with just two points $a = t_0 < t_1 = b$, we deduce that $d(P, Q) \leq l$.

Suppose now that $l = d(P, Q)$ for some Γ; then, for any $t \in [a, b]$, we have

$$\begin{aligned}
d(P, Q) = l &= \text{length } \Gamma|_{[a,t]} + \text{length } \Gamma|_{[t,b]} \\
&\geq d(P, \Gamma(t)) + d(\Gamma(t), Q) \\
&\geq d(P, Q) \qquad\qquad\qquad\qquad \text{by Corollary 2.7.}
\end{aligned}$$

Therefore $d(P, Q) = d(P, \Gamma(t)) + d(\Gamma(t), Q)$ for all t. Applying Corollary 2.7 again, we deduce that $\Gamma(t)$ is on the (shorter) spherical line segment PQ on S for all t, and hence the image of Γ is the spherical line segment. $\qquad\square$

Remark 2.11 So if Γ is a curve of minimum length joining P and Q, it is a spherical line segment. Moreover from the proof of Proposition 2.10, we see that

$$\text{length } \Gamma|_{[0,t]} = d(P, \Gamma(t)),$$

for all t. Thus $d(P, \Gamma(t))$ is strictly increasing as a function of t, which says that the parametrization is *monotonic*.

Summing up the results of this section, we have seen that the spherical metric on S^2 is an intrinsic metric, namely distances are determined by infima of lengths of curves joining given points. Since the distance in this metric between two points is realized as the length of some curve between them (namely, a geodesic line segment), the sphere together with this metric is what we called a geodesic space. Furthermore, if we start instead from the metric on S^2 given by restriction of the Euclidean metric on $\mathbf{R}^3$, then Proposition 2.9 implies that the associated intrinsic metric (defined by means of infima of lengths of curves between points) is precisely the spherical metric.

2.4 Finite groups of isometries

Having now determined the natural metric on S^2, we can ask about its group of isometries $\text{Isom}(S^2)$. We recall from Chapter 1 that an isometry of $\mathbf{R}^3$ which fixes the origin is determined by a matrix in $O(3, \mathbf{R})$. Since such a matrix preserves the standard inner-product, it preserves both the lengths of vectors and the angles between vectors. Since the distance between points of S^2 has been defined to be precisely the angle between the corresponding unit vectors, it is clear that such an isometry of $\mathbf{R}^3$ restricts to an isometry of S^2. Moreover, since any matrix in $O(3)$ is determined by its effect on the standard orthonormal triad of basis vectors in $\mathbf{R}^3$, it is clear that different matrices in $O(3)$ give rise to different isometries of S^2.

We now observe that any isometry $f : S^2 \to S^2$ is of the above form. For this, we note that any such isometry f may be extended to a map $g : \mathbf{R}^3 \to \mathbf{R}^3$ fixing the origin, which for non-zero $\mathbf{x}$ is defined via the recipe

$$g(\mathbf{x}) := \|\mathbf{x}\| f(\mathbf{x}/\|\mathbf{x}\|).$$

Letting $(\, , \,)$ denote the standard inner-product on $\mathbf{R}^3$, we have, for any $\mathbf{x}, \mathbf{y} \in \mathbf{R}^3$, that $(g(\mathbf{x}), g(\mathbf{y})) = (\mathbf{x}, \mathbf{y})$. For $\mathbf{x}, \mathbf{y}$ non-zero, this follows since

$$(g(\mathbf{x}), g(\mathbf{y})) = \|\mathbf{x}\| \, \|\mathbf{y}\| \, \big(f(\mathbf{x}/\|\mathbf{x}\|), f(\mathbf{y}/\|\mathbf{y}\|) \big)$$
$$= \|\mathbf{x}\| \, \|\mathbf{y}\| \, \big(\mathbf{x}/\|\mathbf{x}\|, \mathbf{y}/\|\mathbf{y}\| \big) = (\mathbf{x}, \mathbf{y}),$$

using the property that f preserves the angles between unit vectors and the bilinearity of the inner-product. From this we deduce that g is an isometry of $\mathbf{R}^3$ which fixes the origin, and hence, using Theorem 1.5, is given by a matrix in $O(3)$. In summary therefore, we have seen that Isom(S^2) is naturally identified with the group $O(3, \mathbf{R})$. The results we proved about $O(3, \mathbf{R})$ in Chapter 1 will therefore have precise counterparts for Isom(S^2).

We define a *reflection* of S^2 in a spherical line l (a great circle, say $l = H \cap S^2$ for some plane H passing through the origin) to be the restriction to S^2 of the isometry R_H of $\mathbf{R}^3$, the reflection of $\mathbf{R}^3$ in the hyperplane H. It therefore follows immediately from results in the Euclidean case that any element of Isom(S^2) is the composite of at most three such reflections. We recall in passing that an exactly analogous fact held for the isometries of the Euclidean plane $\mathbf{R}^2$. There is moreover an index two subgroup of Isom(S^2) corresponding to the subgroup $SO(3) \subset O(3)$; these isometries are just the rotations of S^2, and are the composite of two reflections. Since any element of $O(3)$ is of the form $\pm A$, with $A \in SO(3)$, it follows that the group $O(3)$ is isomorphic to $SO(3) \times C_2$.

We now ask about the finite subgroups of Isom(S^2); as above, these will correspond to the finite subgroups of Isom($\mathbf{R}^3$). Conversely, any finite subgroup G of Isom($\mathbf{R}^3$) has a fixed point in $\mathbf{R}^3$, namely

$$\frac{1}{|G|} \sum_{g \in G} g(\mathbf{0}) \in \mathbf{R}^3,$$

and so corresponds to a finite subgroup of Isom(S^2). We saw (Exercise 1.5) that, since any finite subgroup of Isom($\mathbf{R}^2$) has a fixed point, it is either a cyclic or dihedral group. We shall see (Exercise 5.16) that the same statements are true for any finite subgroup of isometries of the hyperbolic plane, although a slightly different argument will be needed to deduce the existence of a fixed point. The group Isom(S^2) = $O(3)$ certainly contains such subgroups, but it is no longer true that any finite subgroup of isometries has a fixed point in S^2, and so there are further subgroups to consider.

We consider first the group of rotations $SO(3)$; by considering rotations of S^2 about the z-axis through angles which are multiples of $2\pi/n$, we see that $SO(3)$ contains copies of the cyclic group C_n. By including also the rotation of S^2 about the x-axis through an angle π, we generate a new subgroup of $SO(3)$ which, for $n > 2$, is isomorphic to the group of symmetries D_{2n} of the regular n-gon, and where for $n = 2$ we have the special case $D_4 = C_2 \times C_2$.

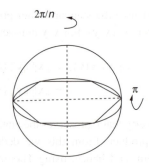

There are however further finite subgroups of $SO(3)$, corresponding to the rotation groups of the regular solids. The tetrahedron has rotation group A_4, and the cube has rotation group S_4 by Exercise 1.15. The octahedron is *dual to* the cube, since by taking the midpoints of the faces of the cube as new vertices, we obtain the octahedron. Thus, any symmetry of the cube gives rise to a symmetry of the octahedron, and vice versa, and so the rotation and full symmetry groups of the cube are the same as those of the octahedron. Standard classical group theory shows that the rotation group of the dodecahedron (with its 12 pentagonal faces and 20 vertices) is A_5, as there are 5 inscribed cubes (each pentagonal face has 5 diagonals, and each one of these diagonals is an edge for one of the cubes), and the rotations act by means of even permutations on these cubes. The icosahedron (with 12 vertices and 20 triangular faces) is dual to the dodecahedron, and so has the same rotation and full symmetry groups. A straightforward, albeit slightly long, argument using elementary group theory [4], shows that we have now accounted for all the finite subgroups of $SO(3)$.

Proposition 2.12   *The finite subgroups of $SO(3)$ are of isomorphism types C_n for $n \geq 1$, D_{2n} for $n \geq 2$, A_4, S_4, A_5, the last three being the rotation groups arising from the regular solids.*   □

We comment now that $-I \in O(3) \setminus SO(3)$, and so if G is a finite subgroup of $SO(3)$, then $H = C_2 \times G$ is a subgroup of $O(3)$ of twice the order, with elements $\pm A$ for $A \in G$. This however may not be the only isomorphism type of subgroups H, of twice the order of G, with $H \cap SO(3) = G$. For instance, although the full symmetry groups of the cube and dodecahedron contain $-I$ and hence are of the type $C_2 \times S_4$ and $C_2 \times A_5$ respectively, the full symmetry group of the tetrahedron is S_4 rather than $C_2 \times A_4$. The complete classification of finite subgroups of $O(3)$ may be found for instance in [4].

Remark 2.13   There is a further reason why these extra finite groups occur for the sphere and not for either the Euclidean or hyperbolic cases. If we have a spherical triangle $\triangle$ with angles $\pi/p, \pi/q$ and π/r, with $r \geq q \geq p \geq 2$, we can consider the subgroup of isometries G generated by the reflections in the sides of the triangle. The theory of *reflection groups* shows that S^2 is *tessellated* by the images of $\triangle$ under the elements of G (cf. [2], Section 9.8). This means that S^2 is covered by the spherical triangles $g(\triangle)$ for $g \in G$, and that the interiors of any two such images are disjoint. Such a tessellation of S^2 gives rise to a rather special type of *geodesic triangulation* (defined in Chapter 3), for which all the triangles are congruent. In particular therefore, the reflection group G is finite.

We see from the Gauss–Bonnet theorem, proved in the next section, that the area of the triangle $\triangle$ is $\pi(1/p + 1/q + 1/r - 1)$, and hence that $1/p + 1/q + 1/r > 1$. The only solutions here are:

- $(p, q, r) = (2, 2, n)$ with $n \geq 2$. The area of $\triangle$ is π/n.
- $(p, q, r) = (2, 3, 3)$. The area of $\triangle$ is $\pi/6$.

- $(p, q, r) = (2, 3, 4)$. The area of $\triangle$ is $\pi/12$.
- $(p, q, r) = (2, 3, 5)$. The area of $\triangle$ is $\pi/30$.

The fact that S^2 (of area 4π) is tessellated by the images of $\triangle$ under G then implies that G has order $4n$, 24, 48 and 120 in these cases. It is then straightforward to check that in the first case G is $C_2 \times D_{2n}$, and in the remaining cases it is the full symmetry group of the tetrahedron, cube and dodecahedron, respectively.

The tessellation of S^2 arising in the first case is clear. For the remaining three cases, let us consider for example the case $(p, q, r) = (2, 3, 4)$. By radial projection from the centre of the cube onto a sphere with the same centre, we obtain a decomposition of S^2 into spherical squares; each edge of the cube gives rise to a plane through the origin in $\mathbf{R}^3$, and hence a spherical line segment on S^2. These spherical squares have angles $2\pi/3$, since three faces meet at each vertex. If we now subdivide each of these spherical squares into eight congruent triangles as shown, each such spherical triangle has angles $\pi/2, \pi/3$ and $\pi/4$, and we obtain a tessellation of S^2 by 48 such triangles. The tessellations in the other two cases follow by a similar construction, starting with the tetrahedron and dodecahedron.

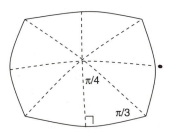

In both the Euclidean and hyperbolic planes, the Euclidean (respectively, hyperbolic) triangles whose angles are all of the form π divided by a positive integer will give rise to interesting reflection groups. The hyperbolic plane is particularly rich in this respect, the resulting tessellations of the hyperbolic plane having been so beautifully exploited in the graphic work of M. C. Escher. Since both the Euclidean and hyperbolic planes have infinite area however, the corresponding reflection groups will be infinite rather than finite.

2.5   Gauss–Bonnet and spherical polygons

In the previous section, we needed the area of a spherical triangle; this formula represents the spherical version of the Gauss–Bonnet theorem. The Euclidean version of Gauss–Bonnet is just the familiar statement that the angles of a Euclidean triangle add up to π.

Proposition 2.14   *If $\triangle$ is a spherical triangle with angles α, β, γ, its area is* $(\alpha + \beta + \gamma) - \pi$.

Proof   A *double lune* with angle $0 < \alpha < \pi$ consists of the two regions on S cut out by two planes passing through two given antipodal points on S, with the angle between the planes being α. In view of the fact that the area of S^2 is 4π, it is clear that the area of the double lune is 4α.

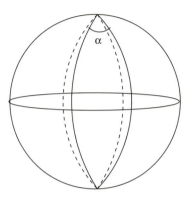

A spherical triangle $\triangle = ABC$ is the intersection of three single lunes — in fact two suffice. Therefore, $\triangle$ and its antipodal triangle $\triangle'$ are in all three of the double lunes (with areas 4α, 4β, 4γ) but any other point of the sphere is in only one of the double lunes, as may be seen with the aid of the diagram below.

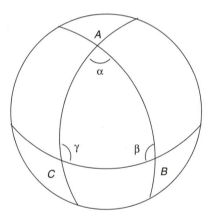

Thus

$$4(\alpha + \beta + \gamma) = 4\pi + 2 \times 2 \times A$$

where 4π is the total area of S^2 and $A = $ area $\triangle = $ area $\triangle'$. Hence the result follows.

$\square$

Remark 2.15

(i) For a spherical triangle, $\alpha + \beta + \gamma > \pi$. In the limit as area $\triangle \to 0$, we obtain $\alpha + \beta + \gamma = \pi$, that is the Euclidean case.

(ii) We may in fact relax our definition of a spherical triangle, by omitting the stipulation that the sides are of length less than π. This is only a minor change, since only one side could have length $\geq \pi$ (otherwise adjacent sides would meet twice, and we would not have a triangle). If however one of the sides has length $\geq \pi$, we can subdivide the triangle into two smaller ones, whose sides have length less than π. Applying Gauss–Bonnet to the two smaller triangles and adding, the area of the original triangle is still

$$\alpha + \beta + \gamma + \pi - 2\pi = \alpha + \beta + \gamma - \pi.$$

We now extend the Gauss–Bonnet to spherical polygons on S^2. Suppose we have a simple closed (spherically) polygonal curve C on S^2, the segments of C being spherical line segments. Let us suppose that the north pole does not lie on C, and we consider the image Γ of C under stereographic projection (as defined in the next section), a simple closed curve in $\mathbf{C}$. By Remark 2.24, the segments of Γ are arcs of certain circles or segments of certain lines.

Applying Propositions 1.17 and 1.19, or rather Remarks 1.18 and 1.20, to Γ, we deduce that the complement of Γ in $\mathbf{C}$ has two components, one bounded and one unbounded. Thus, the complement of C in S^2 also has two path connected components; each of these corresponds to the bounded component in the image of an appropriately chosen stereographic projection. The data of the polygonal curve C and a choice of a connected component of its complement in S^2 determines a *spherical polygon*. The Gauss–Bonnet formula for polygons will play a crucial role in later chapters, in our study of the Euler number and its topological invariance.

A subset A of S^2 is called *convex* if, for any points $P, Q \in A$, there is a unique spherical line segment of minimum length joining P to Q, and this line segment is contained in A. In particular, minimum length spherical line segments in A meet in at most one point. We observe that any open hemisphere is a convex open subset of S^2.

We prove below, in Theorem 2.16, a formula for the area of a spherical n-gon contained in an open hemisphere, namely $\alpha_1 + \cdots + \alpha_n - (n-2)\pi$, where the α_i are the interior angles. This is a combinatorial proof, proceeding by induction on the number of vertices of the polygon. Let us remark first that it is easy to see the validity of the formula for *convex* spherical polygons. This follows immediately from the Gauss–Bonnet formula for spherical triangles, since a convex spherical n-gon may be split into $n-2$ spherical triangles. For clarity, we draw below our spherical polygons as Euclidean ones (for which our arguments also hold).

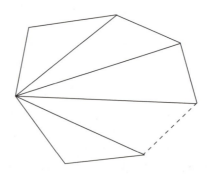

More generally, the formula for area given is easily checked to be *additive*. If Π_1 and Π_2 are spherical polygons, meeting along a common side but otherwise disjoint, then the union is also a spherical polygon Π; if the smaller polygons have n_1 and n_2 sides respectively, the large polygon has $n_1 + n_2 - 2$ sides. The expression given for the area of Π is then just the sum of the corresponding expressions for Π_1 and Π_2. Thus, if the formula for the area of a spherical polygon is true for both Π_1 and Π_2, it will be true for the union.

Theorem 2.16   *If $\Pi \subset S^2$ is a spherical n-gon, contained in some open hemisphere, with interior angles $\alpha_1, \cdots , \alpha_n$, its area is*

$$\alpha_1 + \cdots + \alpha_n - (n - 2)\pi.$$

Proof   The property of the hemisphere we use is that of it being convex, in the sense defined above. We prove the formula by induction on n, the case $n = 3$ following from Gauss–Bonnet. We show that there is always an *internal diagonal* (that is, a spherical line segment joining non-adjacent vertices whose interior is contained in the interior Int Π of Π); this diagonal then divides Π into two polygons, both with strictly less than n sides, and the result follows by induction.

We assume without loss of generality that Π is in the southern hemisphere. We first claim that there is a vertex P which is *locally convex*; this means that for nearby points P', P'' on the boundary of Π either side of P, we obtain a spherical triangle $P'PP''$ which is contained in Π. To see that such a vertex exists, consider a point P of Π at maximum distance from the south pole. For any spherical line segment in the southern hemisphere, the maximum distance from the south pole will be at an end-point; hence P must be a vertex of Π (with interior angle $\alpha < \pi$) and Π must be locally convex at P (see diagram below). Note that both these facts may fail if Π is not contained in an open hemisphere.

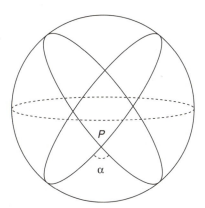

Having found a locally convex vertex, the proof is essentially just a combinatorial argument, and as such it will be valid in other geometries for geodesic n-gons contained in a suitable convex open set. We shall need the following two obvious properties of spherical lines, both of which generalize appropriately for the more general geometries we study later.

(i)   Two distinct spherical lines have at most one point of intersection in any given open hemisphere.

(ii) Two distinct spherical lines have distinct tangents at any given point of intersection (i.e. intersect *transversely*).

Let us consider adjacent vertices P_1, P_2, P_3 of Π, with P_2 locally convex, constructed as above. We may assume therefore that P_2 is not on the spherical line segment $P_1 P_3$. Let l denote the spherical line segment of shortest length joining P_1 to P_3. In this way we obtain a spherical triangle $\triangle = P_1 P_2 P_3$. Since we know the result for spherical triangles contained in a hemisphere, we may assume $n > 3$ and so $\Pi \neq \triangle$. If the interior of l is contained in Int Π, then it is an internal diagonal and there is nothing more to prove; we suppose therefore from now on that this is not the case.

Let Q_t denote the point on the spherical line segment $P_2 P_3$ with

$$d(P_2, Q_t) = t \, d(P_2, P_3),$$

where d here denotes the standard metric on S^2 and where $0 \leq t \leq 1$. We let l_t denote the minimal length spherical line segment $P_1 Q_t$; thus $l_1 = l$. For $t > 0$ sufficiently small, we have that l_t meets the boundary of Π only at the end-points P_1 and Q_t. This follows from property (i) above, since for $t > 0$ sufficiently small, l_t can only meet the two sides of Π that have P_1 as an end-point at the point P_1, and by a continuity argument, the only other side that it can intersect is then $P_2 P_3$ at Q_t. Moreover, by the locally convex property of the vertex P_2, the interior points of l_t near Q_t will be in Int Π, when t is small. Since the interior of l_t is connected, it follows that for $t > 0$ sufficiently small, the interior of l_t is contained in Int Π (since otherwise the interior of l_t may be expressed as a disjoint union of two non-empty open subsets, namely those points which are in Int Π and those which are in the complement of Π).

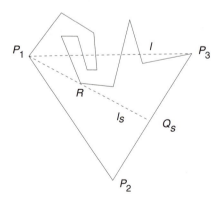

We now consider the supremum s of the values t for which the interior of l_t is contained in Int Π; thus $0 < s \leq 1$. Note that l_s is contained in Π, but will contain points of the boundary $\partial \Pi$ of Π other than P_1 and Q_s. In fact, using the fact that distinct spherical lines can only meet transversely, we see that if l_s contains an interior point of a side of Π, it must contain the whole side; otherwise there would exist $\varepsilon > 0$ such that the interior of l_t intersects $\partial \Pi$ for $s - \varepsilon < t \leq s$, contradicting the assumption that s was a supremum. Thus, in addition to the end-points P_1 and Q_s, we have that $l_s \cap \partial \Pi$ consists of vertices of Π and possibly also some sides of Π; in all cases therefore it contains a vertex R of Π other than P_1 and P_3.

In summary therefore, the interior of the spherical triangle $\triangle' = P_1 P_2 Q_s$ will be contained in Int Π, and s is the largest number for which this is true. It is then clear that the spherical line segment $P_2 R$ has its interior contained in Int $\triangle'$, and hence in Int Π. This then is the internal diagonal we require, dividing Π into two spherical polygons with strictly less than n sides. $\square$

2.6 Möbius geometry

Closely related to spherical geometry is the geometry of Möbius transformations on the extended complex plane $\mathbf{C}_\infty = \mathbf{C} \cup \{\infty\}$, with coordinate ζ say. This connection is provided by the *stereographic projection map*

$$\pi : S^2 \to \mathbf{C}_\infty,$$

defined geometrically by the diagram below. Namely, $\pi(P)$ is the point of intersection of the line through N and P with $\mathbf{C}$, where $\mathbf{C}$ is identified as the plane $z = 0$, and where we define $\pi(N) := \infty$. Clearly, π is a bijection.

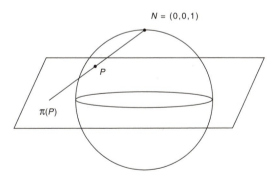

Using the geometry of similar triangles, we can produce an explicit formula for π, namely

$$\pi(x, y, z) = \frac{x + iy}{1 - z}.$$

since in the diagram below $\frac{r}{R} = \frac{1-z}{1}$, and so $R = \frac{r}{1-z}$.

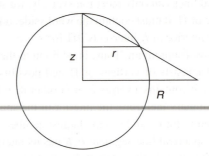

What happens if we project instead from the south pole?

Lemma 2.17   *If $\pi' : S^2 \to \mathbf{C}_\infty$ denotes the stereographic projection from the south pole, then*

$$\pi'(P) = 1/\overline{\pi(P)}$$

for any $P \in S^2$.

Proof   If $P = (x, y, z)$, then $\pi(P) = \frac{x+iy}{1-z}$ and $\pi'(P) = \frac{x+iy}{1+z}$, and hence

$$\overline{\pi(P)}\,\pi'(P) = \frac{x^2 + y^2}{1 - z^2} = 1. \qquad \square$$

Remark 2.18   Thus the map $\pi' \circ \pi^{-1} : \mathbf{C}_\infty \to \mathbf{C}_\infty$ is just inversion in the unit circle, $\zeta \mapsto 1/\bar{\zeta}$.

We shall however consistently adopt the convention that we project from the north pole. For future use, we observe the simple relationship between the images under π of antipodal points.

If $P = (x, y, z) \in S^2$, then $\pi(P) = \zeta = \frac{x+iy}{1-z}$. The antipodal point $-P = (-x, -y, -z)$ has $\pi(-P) = -\frac{x+iy}{1+z}$ and so

$$\pi(P)\,\overline{\pi(-P)} = -\frac{x^2 + y^2}{1 - z^2} = -1.$$

Therefore

$$\boxed{\pi(-P) = -1/\overline{\pi(P)}}$$

Recall now that $\mathbf{C}_\infty$ has the group G of Möbius transformations acting on it. If $A = \left(\begin{smallmatrix} a & b \\ c & d \end{smallmatrix}\right) \in GL(2, \mathbf{C})$, then it defines a Möbius transformation on $\mathbf{C}_\infty$ by

$$\zeta \mapsto \frac{a\zeta + b}{c\zeta + d}.$$

For any $\lambda \in \mathbf{C}^* = \mathbf{C} \setminus \{0\}$, note that λA defines the same Möbius transformation. Conversely, if A_1, A_2 define the same Möbius transformation, then $A_2^{-1} A_1$ defines the identity transformation. It is easily checked that this implies that $A_2^{-1} A_1 = \lambda I$ for some $\lambda \in \mathbf{C}^*$, and hence that $A_1 = \lambda A_2$. Therefore

$$G = PGL(2, \mathbf{C}) := GL(2, \mathbf{C})/\mathbf{C}^*,$$

where the group on the right is obtained by identifying elements of $GL(2, \mathbf{C})$ which are non-zero multiples of each other — formally, it is the quotient of $GL(2, \mathbf{C})$ by the normal subgroup $\mathbf{C}^* I$. We can however always normalize A to have $\det A = 1$. If $\det A_1 = 1 = \det A_2$ and $A_1 = \lambda A_2$, then $\lambda^2 = 1$, and so $\lambda = \pm 1$. Therefore

$$G = PSL(2, \mathbf{C}) := SL(2, \mathbf{C})/\{\pm 1\},$$

where the group on the right is obtained by identifying elements of $SL(2, \mathbf{C})$ which differ only by a sign. The quotient map $SL(2, \mathbf{C}) \to G$ is a surjective group homomorphism which is 2-1; we say that $SL(2, \mathbf{C})$ is a *double cover* of G.

We recall now for future use some elementary facts about Möbius transformations.

(i) The group G of Möbius transformations is generated by elements of the form
- $z \mapsto z + a, \quad$ for $a \in \mathbf{C}$,
- $z \mapsto az, \quad$ for $a \in \mathbf{C}^* = \mathbf{C} \setminus \{0\}$,
- $z \mapsto 1/z$.

(ii) Any circle/straight line in $\mathbf{C}$ is of the form

$$az\bar{z} - \bar{w}z - w\bar{z} + c = 0,$$

for $a, c \in \mathbf{R}$, $w \in \mathbf{C}$ such that $|w|^2 > ac$, and therefore is determined by an indefinite hermitian 2×2 matrix

$$\begin{pmatrix} a & w \\ \bar{w} & c \end{pmatrix}.$$

(iii) From (i) and (ii), it follows that Möbius transformations send circles/straight lines to circles/straight lines.

(iv) Given distinct points $z_1, z_2, z_3 \in \mathbf{C}_\infty$, there exists a unique Möbius transformation T such that $T(z_1) = 0$, $T(z_2) = 1$ and $T(z_3) = \infty$, namely (with appropriate conventions over infinity)

$$T(z) = \frac{z - z_1}{z - z_3} \frac{z_2 - z_3}{z_2 - z_1}.$$

This in particular implies that the action of G is triply transitive, i.e. for any given distinct points $w_1, w_2, w_3 \in \mathbf{C}_\infty$, there exists a (unique) Möbius transformation R such that $R(z_i) = w_i$ for $i = 1, 2, 3$.

(v)   The *cross-ratio* $[z_1, z_2, z_3, z_4]$ of distinct points of $\mathbf{C}_\infty$ is *defined* to be the image of z_4 under the unique map T defined above in (iv).

A comment concerning the invariance of the cross-ratio under Möbius transformations is in order; with the dynamic definition we have adopted here, this invariance is a tautology. Given distinct points z_1, z_2, z_3, z_4 and a Möbius transformation R, there exists a unique Möbius transformation T sending $R(z_1), R(z_2)$ and $R(z_3)$ to 0, 1 and ∞. The composite TR is therefore the unique Möbius transformation sending z_1, z_2 and z_3 to 0, 1 and ∞. Our definition of cross-ratio then immediately implies that

$$[Rz_1, Rz_2, Rz_3, Rz_4] = T(Rz_4) = (TR)z_4 = [z_1, z_2, z_3, z_4].$$

2.7   The double cover of $SO(3)$

On $\mathbf{C}_\infty$, we have an action of the group $PSU(2) = SU(2)/\{\pm 1\}$, which may be identified as the group of Möbius transformations defined by elements of $SU(2) \subset SL(2, \mathbf{C})$. Recall that $SU(2)$ consists of matrices of the form $\left(\begin{smallmatrix} a & -b \\ \bar{b} & \bar{a} \end{smallmatrix}\right)$ with $|a|^2 + |b|^2 = 1$. On S^2 we have the rotations $SO(3)$, an index two subgroup of the full isometry group $O(3)$. The purpose of this section is to show that, via the stereographic projection map π, the group $SO(3)$ is identified isomorphically with the group $PSU(2)$. In particular, we have a surjective homomorphism of groups $SU(2) \to SO(3)$, which is 2-1 map.

Theorem 2.19   *Via the map π, every rotation of S^2 corresponds to a Möbius transformation of $\mathbf{C}_\infty$ in $PSU(2)$.*

Proof   *Step 1*: The rotation $r(z, \theta)$ about the z-axis $\mathbf{R}(0, 0, 1)^t$, through an angle θ (clockwise), corresponds under π to the Möbius transformation $\zeta \mapsto e^{i\theta}\zeta$, defined by the matrix

$$\begin{pmatrix} e^{i\theta/2} & 0 \\ 0 & e^{-i\theta/2} \end{pmatrix} \in SU(2).$$

Step 2: Now consider the rotation $r(y, \pi/2)$ given by the matrix

$$\begin{pmatrix} 0 & 0 & 1 \\ 0 & 1 & 0 \\ -1 & 0 & 0 \end{pmatrix}.$$

In $\mathbf{C}_\infty$, the corresponding map is

$$\frac{x + iy}{1 - z} = \zeta \mapsto \zeta' = \frac{z + iy}{1 + x}.$$

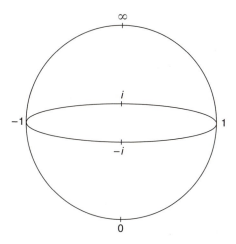

Labelling points on S^2 by the corresponding elements of $\mathbf{C}_\infty$, we see that, if this map does correspond to a Möbius transformation, it has to be $\zeta' = \frac{\zeta-1}{\zeta+1}$ (since a Möbius transformation is determined by its action on a triple of points, and $-1 \mapsto \infty, 1 \mapsto 0$ and $i \mapsto i$). We now check that it is this transformation:

$$\frac{\zeta - 1}{\zeta + 1} = \frac{x + iy - 1 + z}{x + iy + 1 - z}$$
$$= \frac{x - 1 + z + iy}{x + 1 - (z - iy)}$$
$$= \frac{(z + iy)(x - 1 + z + iy)}{(x + 1)(z + iy) + x^2 - 1}$$
$$= \frac{(z + iy)(x - 1 + z + iy)}{(x + 1)(z + iy + x - 1)} = \zeta'$$

as required. We observe that the Möbius transformation is defined by the matrix

$$\frac{1}{\sqrt{2}} \begin{pmatrix} 1 & -1 \\ 1 & 1 \end{pmatrix} \in SU(2).$$

Step 3: We claim that $SO(3)$ is generated by $r(y, \pi/2)$ and rotations of the form $r(z, \theta)$, $0 \leq \theta < 2\pi$. First observe that, for any angle ϕ, the rotation

$$r(x, \phi) = r(y, \pi/2)\, r(z, \phi)\, r(y, -\pi/2)$$

is a composite of these generators. Also, for any $\mathbf{v} \in S^2$, there exist ϕ, ψ such that $g = r(z, \psi)\, r(x, \phi)$ sends $\mathbf{v}$ to $(1, 0, 0)^t$ (we rotate $\mathbf{v}$ first to the horizontal and then to $(1, 0, 0)^t$), and so this g is also a product of generators of the type claimed. A rotation about $\mathbf{v}$ through a (clockwise) angle θ may be written as

$$r(\mathbf{v}, \theta) = g^{-1} r(x, \theta) g,$$

which implies that any rotation $r(\mathbf{v}, \theta)$ can be written as the product of elements of the type claimed.

Step 4: Hence, via π, any rotation of S^2 gives rise to a product of the Möbius transformations of $\mathbf{C}_\infty$ corresponding to these generators, and these are all defined by matrices in $SU(2)$. The claimed result has now been proved.   □

Theorem 2.20   *The group of rotations $SO(3)$ acting on S^2 corresponds isomorphically with the subgroup $PSU(2) = SU(2)/\{\pm1\}$ of Möbius transformations acting on $\mathbf{C}_\infty$.*

Proof   In the previous theorem, we produced an injective homomorphism from the rotation group $SO(3)$ to the subgroup $PSU(2)$ of the group of Möbius transformations. We now need to show that this map is surjective. Given $g \in PSU(2)$ a Möbius transformation, say

$$g(z) = \frac{az - b}{bz + \bar{a}},$$

we need to show that it corresponds under π to a rotation of S^2. We suppose first that $g(0) = 0$; then $b = 0$ and $a\bar{a} = 1$, and so $a = e^{i\theta/2}$ for some θ, and hence g corresponds to $r(z, \theta)$.

In general, we suppose $g(0) = w \in \mathbf{C}_\infty$ and let $Q \in S^2$ be such that $\pi(Q) = w$. We choose a rotation A of S^2 with $A(Q) = (0, 0, -1)^t$ and let α be the corresponding element of $PSU(2)$; therefore $\alpha(w) = 0$. We note that $\alpha \circ g$ fixes 0, and hence from the previous case corresponds to a rotation $B = r(z, \theta)$. Thus g corresponds to the composite rotation $A^{-1}B$.   □

Corollary 2.21   *The isometries of S^2 which are not rotations correspond under stereographic projection precisely to the transformations of $\mathbf{C}_\infty$ of the form*

$$z \mapsto \frac{a\bar{z} - b}{b\bar{z} + \bar{a}},$$

with $|a|^2 + |b|^2 = 1$.

Proof   If R denotes the reflection of S^2 in the xz-plane, any isometry of S^2 which is not a rotation can be written in the form AR, where $A \in SO(3)$. Since R corresponds on $\mathbf{C}$ to complex conjugation, the claim follows from the theorem.   □

So we have seen that there exists a 2-1 map

$$SU(2) \rightarrow PSU(2) \cong SO(3).$$

This map is usually produced using quaternions; see for instance Chapter 8 of [10] for details.

This is the reason why there exists a non-closed path of transformations in $SU(2)$ going from I to $-I$, corresponding to a closed path in $SO(3)$ starting and ending at

$$\begin{pmatrix} 1 & 0 & 0 \\ 0 & 1 & 0 \\ 0 & 0 & 1 \end{pmatrix}.$$

This last fact may be demonstrated experimentally as follows. Rest a plate on the upturned fingers of your hand. Now twist your arm, thus also twisting the plate. Theoretically, one can do this, keeping the centre of the plate fixed, and so the position of the plate at any given time is represented by a rotation in $SO(3)$, and the whole operation is, from the point of view of the plate, described by a path in $SO(3)$. One finds however that on the first occasion that the plate returns to its original position, the arm will still be twisted. If one continues to twist the plate and arm, one discovers (on the second occasion that the plate returns to its original position) that the arm also returns to its original state of being untwisted. Thus, the history of the position of the plate and the twistedness of your arm could be represented by a simple closed path in $SU(2)$, the position of only the plate corresponding to the projection onto $SO(3)$. With the experiment suggested, the projected path in $SO(3)$ will already have returned to its starting point halfway through the experiment. Those readers who have difficulties visualising this experiment might wish to consult page 166 of [3], where there is a series of photographs illustrating it, with a mug rather than a plate.

Remark 2.22

(i) Since $SU(2)$ consists of matrices of the form $\left(\begin{smallmatrix} a & -b \\ b & \bar{a} \end{smallmatrix} \right)$ with $|a|^2 + |b|^2 = 1$, geometrically it is $S^3 \subset \mathbf{R}^4$. It is well known however that the space S^3 has no non-trivial covers; this property is equivalent to the property that S^3 is *simply connected*, namely that any closed path on S^3 may be continuously shrunk down to a point.

(ii) Corresponding to the finite subgroups of $SO(3)$, namely cyclic, dihedral, and the rotation groups of the tetrahedron, cube and dodecahedron, there are finite subgroups of $SU(2)$ of twice the order. Corresponding to a subgroup C_n of $SO(3)$, we clearly obtain a cyclic subgroup C_{2n} by Step 1 from Theorem 2.19. The subgroup of $SU(2)$ corresponding to a dihedral group D_{2n} in $SO(3)$ is called the *dicyclic* group; it contains a subgroup C_{2n}, but the order 2 subgroup of D_{2n} generated by the rotation of S^2 through an angle π about the x-axis lifts to a cyclic subgroup of order 4. In the case of $D_4 = C_2 \times C_2$, the dicyclic group we obtain is just the well-known quaternion group of order eight. The subgroups of $SU(2)$ of orders 24, 48 and 120, corresponding to the three rotation groups of regular solids, are usually called the *binary tetrahedral, binary octahedral* and *binary icosahedral* groups.

2.8 Circles on S²

Given an arbitrary point P on S^2, and $0 \le \rho < \pi$, we may consider the locus of points on S^2 whose spherical distance from P is ρ. This is what we mean by a *circle* in spherical geometry. We may always rotate the sphere so that the point P is at the north pole, as shown in the figure below.

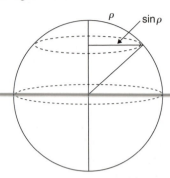

Hence it is clear that the circle is also a Euclidean circle, of radius $\sin \rho$, and that it is the intersection of a plane with S^2. Conversely, any plane whose intersection with S^2 consists of more than one point, cuts out a circle. Recall that the great circles just correspond to the planes passing though the origin. In Exercise 2.6, we calculate the area of such a circle to be

$$2\pi (1 - \cos \rho) = 4\pi \sin^2(\rho/2),$$

which is always less than the area $\pi \rho^2$ from the Euclidean case, and for small ρ may be expanded as

$$\pi \rho^2 \left(1 - \frac{1}{12}\rho^2 + O(\rho^4) \right).$$

We observe that a circle on S^2 which passes through the north pole is cut out by a plane H in $\mathbf{R}^3$ which passes through the north pole, and that under stereographic projection this projects to a line in $\mathbf{C}$, namely the intersection of H with the complex plane (positioned equatorially). Conversely, any line l in $\mathbf{C}$ determines a plane H passing through the north pole, and hence, under stereographic projection, to a circle in S^2 passing through the north pole.

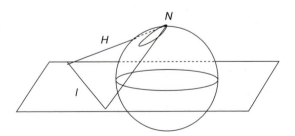

Proposition 2.23   *Under stereographic projection, the circles on S^2 not passing through the north pole correspond to the circles in* $\mathbf{C}$.

Proof   Let us first show that any circle C on S^2, not passing through the north pole, stereographically projects to a circle in $\mathbf{C}$. We choose a rotation A of S^2 which sends the centre of C to the north pole. If C' denotes the image of C under A, by symmetry it is then clear that the stereographic projection of C' is a circle Γ in $\mathbf{C}$. Let α denote the Möbius transformation of $\mathbf{C}_\infty$ corresponding to the rotation A, using Theorem 2.19. Since C is the image of C' under A^{-1}, the stereographic projection of C is the image of the circle Γ under the Möbius transformation α^{-1}, and hence is a circle or straight line. Since C is assumed not to pass through the north pole, it does not correspond to a straight line.

Conversely, suppose Γ is a circle in $\mathbf{C}$, and choose any three distinct points of S^2 whose images under stereographic projection lie on Γ. These three points determine a plane in $\mathbf{R}^3$, and hence a circle C on S^2. The previous argument shows that the stereographic image of C is a circle or straight line Γ', whose intersection with our original circle Γ consists of at least three distinct points. Therefore we deduce that $\Gamma = \Gamma'$, and that it is the stereographic projection of the circle C. Since Γ was assumed to be a circle, we note that C does not pass through the north pole. $\square$

Remark 2.24   The great circles on S^2 will be of three types. Those passing through the north (and hence also south) pole will correspond to lines through the origin in $\mathbf{C}$. The equator is also a special case, in that it corresponds to the unit circle in $\mathbf{C}$. Any other great circle on S^2 projects to a circle in $\mathbf{C}$ which intersects the unit circle at precisely two points, one of which is the negative of the other (corresponding to the antipodal points where our given great circle on S^2 intersects the equator). Conversely, suppose that we have a circle in $\mathbf{C}$ which intersects the unit circle at precisely two points, one of which is the negative of the other. By Proposition 2.23, this corresponds to a circle on S^2, distinct from the equator and not passing through the north pole, and our assumption implies that this circle intersects the equator at antipodal points; it is therefore a great circle.

Exercises

2.1   Given distinct points P, Q on the sphere S^2, use the result from Lemma 1.6 to show that the locus of points on the sphere equidistant from P and Q forms a great circle.

2.2   Given a spherical line l on the sphere S^2 and a point P not on l, show that there is a spherical line l' passing through P and intersecting l at right-angles. Prove that the minimum distance $d(P, Q)$ of P from a point Q on l is attained at one of the two points of intersection of l with l', and that l' is unique if this minimum distance is less than $\pi/2$.

2.3   Show that a spherical triangle (with side-lengths less than π) must be contained in some open hemisphere of S^2.

2.4   Given distinct spherical lines l_1, l_2, defining reflections R_1, R_2 of the sphere, describe geometrically the composite $R_1 R_2$. State and prove the result for spherical triangles which corresponds to the result for Euclidean triangles described in Exercise 1.4.

2.5 Two spherical triangles $\triangle_1, \triangle_2$ on a sphere S^2 are said to be *congruent* if there is an isometry of S^2 that takes $\triangle_1$ to $\triangle_2$. Show that $\triangle_1, \triangle_2$ are congruent if and only if they have equal angles. What other conditions for congruence can you find?

2.6 Given a circle on S^2 of radius ρ in the spherical metric, show that its area is $2\pi(1 - \cos\rho)$.

2.7 Assuming the existence of the regular dodecahedron, demonstrate the existence of a tessellation of S^2 by spherical triangles whose angles are $\pi/2$, $\pi/3$ and $\pi/5$. In a similar way, demonstrate the existence of this tessellation directly in terms of the regular icosahedron.

2.8 Let $\triangle^c$ denote the complement of a spherical triangle $\triangle$ on S^2, and let α, β, γ denote the interior angles for $\triangle^c$ (i.e. the *exterior angles* for $\triangle$). Prove that

$$\text{area } \triangle^c = \alpha + \beta + \gamma - \pi.$$

2.9 Suppose that Γ is a circle or straight line in the complex plane, containing distinct points z_1, z_2, z_3. If T denotes the unique Möbius transformation of $\mathbf{C}_\infty$ sending the points z_1, z_2, z_3 to respectively $0, 1, \infty$, show that any further (distinct) point z_4 lies on Γ if and only if $T(z_4)$ is real. Deduce that four distinct points in the complex plane lie on a circle or straight line if and only if their cross-ratio is real.

2.10 Show that any two distinct circles on the sphere meet in at most two points.

2.11 Let $u, v \in \mathbf{C}_\infty$ correspond under stereographic projection to points P, Q on S^2, and let d denote the spherical distance from P to Q on S^2. Show that $-\tan^2 \frac{1}{2}d$ is the cross-ratio of the points $u, v, -1/\bar{u}, -1/\bar{v}$, taken in an appropriate order (which you should specify).

2.12 If two spherical line segments on S^2 meet at a point P (other than the north pole) at an angle θ, show that, under the stereographic projection map π, the corresponding segments of circles or lines in $\mathbf{C}$ meet at $\pi(P)$, with the same angle and the same orientation. [You may assume that Möbius transformations preserve angles and their orientations; this in fact follows from our discussion in Section 4.1 of complex analytic functions in one complex variable.]

2.13 For every spherical triangle $\triangle = ABC$, show that $a < b + c$, $b < c + a$, $c < a + b$ and $a + b + c < 2\pi$. Conversely, show that, for any three positive numbers a, b, c less than π satisfying the above conditions, we have $\cos(b + c) < \cos a < \cos(b - c)$, and that there is a spherical triangle (unique up to isometries of S^2) with those sides.

2.14 Show that any Möbius transformation T on $\mathbf{C}_\infty$ which is not the identity has one or two fixed points. Show that the Möbius transformation corresponding (under the stereographic projection map) to a rotation of S^2 through a non-zero angle has exactly two fixed points z_1 and z_2, where $z_2 = -1/\bar{z}_1$. If now T is a Möbius transformation with two fixed points z_1 and z_2 satisfying $z_2 = -1/\bar{z}_1$, prove that *either* T corresponds to a rotation of S^2, *or* one of the fixed points, say z_1, is an *attractive* fixed point, i.e. for $z \neq z_2$, we have that $T^n z \to z_1$ as $n \to \infty$.

2.15 For any finite set of points in the Euclidean plane, show that there is a unique circle of minimum radius which encloses them (some will of course lie on the circle). If now we consider the finite set of points of S^2 corresponding to the vertices of a regular solid, show that there is no such *unique* spherical circle which encloses them all. [Hint: Show that the corresponding group of symmetries has no fixed point on S^2.]

2.16   Prove that the formula for the area of a spherical polygon remains true for polygons not necessarily contained in an open hemisphere. [Hint: A limiting argument verifies the formula for spherical polygons contained in a closed hemisphere. For a general polygon in S^2, use the equator to decompose it into a finite union of polygons, each of which is contained in one of the two closed hemispheres.]

3 Triangulations and Euler numbers

3.1 Geometry of the torus

Before starting on the main topics of this chapter, we introduce another geometry, that of the locally Euclidean torus.

Definition 3.1   The *torus* $T = T^2$ may be defined most easily as follows. As a set it is just $\mathbf{R}^2/\mathbf{Z}^2$, whose points are represented by $(x, y) \in \mathbf{R}^2$ modulo the equivalence relation $(x_1, y_1) \sim (x_2, y_2) \iff x_2 - x_1 \in \mathbf{Z}$ and $y_2 - y_1 \in \mathbf{Z}$. We shall denote the quotient map by $\varphi : \mathbf{R}^2 \to T$.

If Q is any closed square in $\mathbf{R}^2$ with vertices at the real points (p, q), $(p + 1, q)$, $(p, q + 1)$, $(p + 1, q + 1)$, then T is also given by identifying sides of Q in pairs as shown; we shall call Q a *fundamental square* for T.

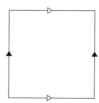

Using our first definition of T, a distance function d may be defined on T as follows:

$$d(P_1, P_2) = \min\{\|\mathbf{x}_1 - \mathbf{x}_2\| : \mathbf{x}_1, \mathbf{x}_2 \in \mathbf{R}^2 \text{ representing } P_1, \text{ respectively } P_2\}.$$

An easy check verifies that d is a metric — in fact T has far more structure than this, as it is a smooth surface in the sense of Chapter 8.

On the interior $\operatorname{Int} Q$ of a square of the above form, the natural map $\varphi|_{\operatorname{Int} Q} : \operatorname{Int} Q \to T$ is a bijection onto an open subset W of T (the complement of two 'circles' S^1 meeting at a point, one of these circles corresponding to a horizontal side of Q, and one to a vertical side).

Given $P \in \operatorname{Int} Q$, the restriction of φ to a small enough open ball B around P is an isometry — thus $\varphi|_{\operatorname{Int} Q} : \operatorname{Int} Q \to W$ is a *homeomorphism* (both $\varphi|_{\operatorname{Int} Q}$ on $\operatorname{Int} Q$ and its inverse on W are continuous).

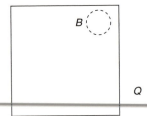

If however Q_1 is a closed square in $\mathbf{R}^2$ with vertices (p, q), $(p+1/2, q)$, $(p, q+1/2)$, $(p+1/2, q+1/2)$, a square of side-length $1/2$, one checks easily (Exercise 3.3) that the restriction of φ to Int Q_1 is an *isometry* onto its image. Moreover, the image of Int Q_1 is a *convex* open subset of T, in the sense of Definition 3.3.

Remark 3.2   The metric d we have put on the torus coincides with the Euclidean metric on small enough discs; it is therefore called *locally Euclidean*, and for many purposes it is the most natural metric to choose. One can of course scale distances differently in the two directions, or alternatively express the torus as the quotient of $\mathbf{R}^2$ by a *rectangular* lattice. More generally, one can consider the quotient of $\mathbf{R}^2$ by an arbitrary lattice (an Abelian subgroup of $\mathbf{R}^2$ generated by two vectors which are linearly independent over the reals), and then the Euclidean metric on $\mathbf{R}^2$ induces a general locally Euclidean metric on the torus. The geometry of all these spaces is very similar, although one observes for instance that the group of isometries fixing a given point is $C_2 \times C_2$ for the rectangular lattice, and the dihedral group D_8 for the square lattice (see Exercise 3.4). For definiteness, we shall always restrict ourselves to the case of the torus T being given as the quotient of $\mathbf{R}^2$ by a unit square lattice, with locally Euclidean metric induced from the metric on $\mathbf{R}^2$.

One should comment however that there are other, very different, metrics which can be defined on T. The torus may be embedded in $\mathbf{R}^3$, for example via the embedding of T induced from $\sigma : \mathbf{R}^2 \to \mathbf{R}^3$ given by

$$\sigma(u, v) = ((2 + \cos u) \cos v, (2 + \cos u) \sin v, \sin u).$$

A metric may be defined on T by considering lengths of curves on the embedded torus; equivalently, this is just the intrinsic metric obtained on T, when one starts with the metric on the embedded torus given by restricting the Euclidean metric from $\mathbf{R}^3$ and applies the recipe from Section 1.4.

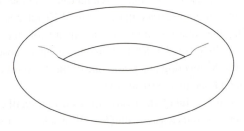

The geometry of the above metric, arising when T is considered as an embedded surface in $\mathbf{R}^3$ is very different from that of the locally Euclidean metric. Such embedded surfaces will be studied systematically in Chapter 6; for the present we shall just concentrate on the locally Euclidean case, as defined above.

We now ask which curves on T should be considered as (geodesic) lines in the geometry. Given two points P_1 and P_2 in T, we can by definition represent them by points $\mathbf{x}_1$ and $\mathbf{x}_2$ in $\mathbf{R}^2$, for which $d(P_1, P_2)$ is just the Euclidean distance between $\mathbf{x}_1$ and $\mathbf{x}_2$. The straight line segment is the unique curve of minimal length between $\mathbf{x}_1$ and $\mathbf{x}_2$, and its image on T is easily checked to be a curve of minimal length between P_1 and P_2. This then motivates the following definition, which will be set in a more general context once we reach Chapter 7. We should however first extend our definition of a *curve* on a metric space X to include continuous functions $\gamma : I \to X$, for I any real interval (which may be finite or infinite, and which may be open or closed at its end-points).

Definition 3.3 A curve $\gamma : I \to T$ on T will be called a *geodesic line* if, under the local identification of an open subset of T with the interior of a fundamental square in $\mathbf{R}^2$, the curve is locally a line segment in the plane. If I is a finite closed real interval, the curve γ is called a *geodesic line segment*. A curve $\gamma : [a, b] \to T$ will be called *polygonal* if it is the concatenation of a finite number of geodesic line segments.

Similarly to the case of the sphere, a subset A of T is called *convex* if for any two points $P, Q \in A$, there is a unique geodesic line segment of minimal length in T joining them, and this geodesic line segment is contained in A. For instance any open ball in T of radius less than $1/4$ is contained in $\varphi(\text{Int } Q_1)$, for some square Q_1 in $\mathbf{R}^2$ with vertices (p, q), $(p + 1/2, q)$, $(p, q + 1/2)$, $(p + 1/2, q + 1/2)$. It is then isometric to an open ball in $\mathbf{R}^2$, and is a convex open subset of T (Exercise 3.3).

As in the case of S^2, we have some geodesic lines which are simple closed curves of T, and so a geodesic line segment between two points of T may therefore not be the curve segment of minimum length. Moreover, there are closed geodesic lines of different types, according to how many times the closed geodesic line winds round the torus in the two directions. We illustrate below a geodesic line which winds round the torus four times in one direction and only once in the other direction.

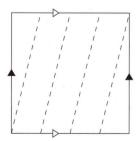

In fact, it is easily seen that a line in $\mathbf{R}^2$ with rational gradient p/q, with p, q *coprime* positive integers, determines a closed geodesic line which winds round the torus p

times in one direction and q times in the other. There are however further new features which occur. If for instance one considers a line in $\mathbf{R}^2$ whose gradient is *irrational*, then its image may be checked to be a non-closed geodesic line on T, of infinite length, whose corresponding point set is *dense* in T, that is the closure of the geodesic line is all of T.

It will also be necessary to modify our definition of what we mean by a polygon on T, since a simple closed (polygonal) curve will not in general subdivide the torus into two path connected components. Representing the torus as a square with opposite sides identified, the line segment joining the midpoints of a pair of opposite sides will clearly define a closed geodesic line, but its complement will only have one connected component. For more complicated spaces S, one may have a simple closed curve which does subdivide S into two path connected components, but neither of these components is topologically a disc. Such an example is provided by the two-holed torus, illustrated below, with a disconnecting curve of the form shown.

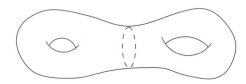

The definition of a polygon that we adopt below can be shown to be equivalent to the simple closed polygonal curve having an inside and an outside, with the *inside* being homeomorphic to the unit disc. For the sphere S^2, it is plainly equivalent to our definition from the previous chapter. The definition also generalizes in an obvious way to more general surfaces, as studied in subsequent chapters, once we know what we mean by a geodesic (line) segment.

Definition 3.4   A (geodesic) *polygon* in $X = S^2$ or T is defined as follows. Suppose that we have a simple closed polygonal curve Γ in X with the following two properties:

(i)   There is a homeomorphism $f : U \to V$ from an open subset of $\mathbf{R}^2$ onto an open subset V of X containing Γ.

(ii)   The simple closed curve $f^{-1} \circ \Gamma$ in $\mathbf{R}^2$ has complement consisting of two connected components, the bounded component being contained in U.

The *open polygon* in X is defined to be the image of this bounded component under f, and the (closed) *polygon* just the closure of this. The polygon therefore has boundary given by the simple closed polygonal curve Γ.

It follows from Chapter 2 that a simple closed polygonal curve on S^2 gives rise to two possible polygons, depending on a choice for the inside of the curve.

With the notation as in the above definition, suppose that we have a simple closed polygonal curve Γ on the locally Euclidean torus T, which is contained in

$W = \varphi(\text{Int } Q)$, for Q some fundamental unit square $\mathbf{R}^2$. Let $\tilde{\Gamma}$ denote the lift of Γ to Int Q, i.e. the unique curve in Int Q with $\Gamma = \varphi \circ \tilde{\Gamma}$. By Proposition 1.19, the complement of $\tilde{\Gamma}$ in $\mathbf{R}^2$ has two components, one of which contains the complement of Int Q and one of which is a plane polygon contained in Int Q. The latter defines a polygon on the torus, contained in W, in the sense defined above. Conversely, since $\varphi|_{\text{Int } Q} : \text{Int } Q \to W$ is a homeomorphism, any polygon on T which is contained in W and has boundary Γ lifts uniquely to a polygon in Int Q with boundary $\tilde{\Gamma}$, and hence is of the above form. Since the projection φ is locally an isometry, we may define the interior angles of the polygon in W to be the interior angles of the lifted Euclidean polygon in Int Q.

In the previous chapter, we proved the Gauss–Bonnet formula for the area of a spherical polygon. The Gauss–Bonnet formula for Euclidean polygons in $\mathbf{R}^2$ says that the sum of the interior angles of an n-gon is $(n-2)\pi$. This Euclidean version of Gauss–Bonnet either follows directly, using our combinatorial proof of Theorem 2.16 to reduce down to the case of triangles, or, alternatively, we may obtain the result as a limit of Theorem 2.16 when the area of the polygon goes to zero. For polygons on the locally Euclidean torus T, the same formula holds, but we shall only need it for polygons contained in open sets $W = \varphi(\text{Int } Q)$, for some fundamental unit square $Q \subset \mathbf{R}^2$; for such polygons, it reduces (via the previous paragraph) to the Euclidean version.

Lemma 3.5 *For n-sided polygons on T contained in $W = \varphi(\text{Int } Q)$, for some fundamental unit square $Q \subset \mathbf{R}^2$, the sum of the interior angles is $(n-2)\pi$.*

3.2 Triangulations

Definition 3.6 A *topological triangle* on a metric space X is defined to be the image of a closed triangle R in $\mathbf{R}^2$, under a homeomorphism $f : U \to V$ from an ϵ-neighbourhood U of R in $\mathbf{R}^2$ (for $\epsilon > 0$ sufficiently small) to an open subset V of X. The boundary of a topological triangle is therefore a simple closed curve in X, and the interior of the topological triangle is homeomorphic to the interior of R, which in turn is homeomorphic to the open disc, by Exercise 1.6.

Remark 3.7 By definition, the ϵ-neighbourhood U of R in $\mathbf{R}^2$ consists of all points $\mathbf{z} \in \mathbf{R}^2$ whose *distance* from R

$$d(\mathbf{z}, R) := \inf\{\|\mathbf{z} - \mathbf{w}\| : \mathbf{w} \in R\}$$

is less than ϵ. The boundary of such an ϵ-neighbourhood U of R is a simple closed curve C in $\mathbf{R}^2$ consisting of three straight line segments and three circular arcs, and so by Proposition 1.17, the complement of U in $\mathbf{R}^2$ is connected (the complement of C has two components, the bounded component being U).

We remark that a spherical triangle on S^2 is an example of a topological triangle, as may be seen most easily by radial projection from the centre of the sphere (Exercise 3.5).

Definition 3.8   A (topological) *triangulation* of a compact metric space X consists of a finite collection of topological triangles whose union is all of X, with the following additional properties:

- Two triangles are either disjoint, or their intersection is either a common vertex or a common edge.
- Each edge is an edge of exactly two triangles.

The *Euler number* (or *Euler characteristic*) of the triangulation is defined to be

$$e = F - E + V,$$

where $F = \#$ triangles, $E = \#$ edges and $V = \#$ vertices.

Remark   To avoid the risk of confusion, an immediate remark is in order here. With the definition of a triangulation we have given here, any space with a triangulation has to be two-dimensional (since it will be locally homeomorphic to an open subset of $\mathbf{R}^2$). By extending the concept of a topological triangle to *simplices* of higher dimensions, we can give an alternative definition of a triangulation, which will be useful in all dimensions. In this course, we shall only consider triangulations of surfaces, and so the two definitions will be the same.

For both the sphere and the torus, we observe below that triangulations exist, and we show that the Euler number does not depend on the choice of triangulation. This latter fact is usually proved by developing the theory of homology groups in Algebraic Topology (as for instance in [7], Chapter IX), but we prove it in this chapter for S^2 and T by purely elementary means. The proof we give below for the sphere and torus will be generalized in Chapter 8 to work on arbitrary compact surfaces. The Euler numbers of both S^2 and T may therefore be readily calculated, just by writing down one triangulation in each case, and we see that $e(S^2) = 2$ and $e(T) = 0$.

Example

(i)   There is a triangulation on S^2 consisting of eight spherical triangles whose angles are all $\pi/2$. This has $F = 8$, $E = 12$, $V = 6$, and so $e = 2$.

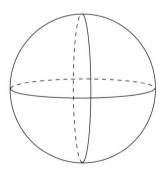

(ii)   There is a triangulation on T as illustrated below (on a fundamental square).

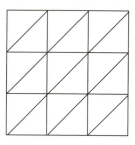

This has $F = 18$, $E = 27$, $V = 9$, and so $e = 0$.

Both these are examples of *geodesic triangulations*, where each side of each triangle is a spherical line segment in the case of S^2, and, in the case of the torus, corresponds to a line segment in the interior Int Q for some unit fundamental square $Q \subset \mathbf{R}^2$. The reader should convince herself why, with the strict definition we gave above, the following decomposition is not a triangulation of T.

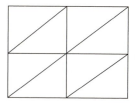

We now describe a construction which enables us to subdivide a triangulation on a metric space X, under which the topological triangles are subdivided into smaller triangles, but for which the Euler number is unchanged. This construction will be used in our proof that the Euler number of a compact surface does not depend on the choice of topological triangulation.

Construction 3.9   Given any triangle R in $\mathbf{R}^2$, and a choice of interior point for each of the sides, we can subdivide R into four smaller triangles as shown. We can moreover iterate this construction, as illustrated.

 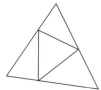 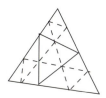

Suppose now we are given a triangulation of a metric space (X, d), and a choice of interior point for each of the edges of the triangulation. For each topological triangle $\hat{R} \subset X$, where by definition there exists a homeomorphism f from some ϵ-neighbourhood of a Euclidean triangle $R \subset \mathbf{R}^2$ to an open neighbourhood of

$\hat{R}$ in X, the interior points chosen on the edges of $\hat{R}$ determine interior points on the sides of R. We make the subdivision of R specified as above by these points, thus obtaining four smaller topological triangles, namely the images under f of each of the four sub-triangles of R. Subdividing all the triangles of the triangulation in this way, we obtain a new triangulation of X, with four times as many triangles. An elementary check confirms that this subdivision leaves the Euler number unchanged.

Definition 3.10   Given a metric space (X, d), the *diameter* of a subset $Y \subset X$ is defined to be $\sup\{d(P, P') : P, P' \in Y\}$.

The following result enables us to reduce down to the consideration of triangulations, all of whose triangles are suitably small.

Proposition 3.11   *Given a metric space X and real number $\varepsilon > 0$, we may (by repeated application of the above construction) replace any given triangulation on X by one with the same Euler number, but all of whose triangles have diameter less than ε.*

Proof We concentrate on a topological triangle $\hat{R}$ of our given triangulation; by definition, this is the image of a Euclidean triangle R under a homeomorphism $f : R \to \hat{R}$. Since R is compact, f is uniformly continuous on R (Lemma 1.13), and so there exists $\delta > 0$ such that, if $\mathbf{x}, \mathbf{y} \in R$ with $\|\mathbf{x} - \mathbf{y}\| < \delta$, then $d(f(\mathbf{x}), f(\mathbf{y})) < \varepsilon$. We let l denote the maximum of the side-lengths of R. We now subdivide R at the midpoints of its edges, hence determining interior points of the corresponding edges of $\hat{R}$. By choosing interior points of the other edges of the triangulation arbitrarily and subdividing as in Construction 3.9, we obtain a new triangulation under which each topological triangle has been replaced by four smaller topological triangles. By repeating this operation on R of subdivision at the midpoints of sides m times, and suitably extending to the rest of the triangulation as in Construction 3.9, we obtain a subdivision of the original triangulation under which each topological triangle has been replaced by 4^m smaller ones. The Euclidean triangle R has been subdivided into 4^m triangles, each with side-lengths 2^{-m} times those of R, and hence of diameter $2^{-m}l$ (see Exercise 5.15). If m is chosen so that $2^{-m}l < \delta$, the topological triangle $\hat{R}$ has been subdivided into 4^m topological triangles in X, each of diameter less than ε.

We consider now another topological triangle $\hat{R}'$ of the original triangulation, where $\hat{R}'$ is the image of a Euclidean triangle R' in $\mathbf{R}^2$ under a homeomorphism h. As before, there exists $\delta' > 0$ such that, if $\mathbf{x}, \mathbf{y} \in R'$ with $\|\mathbf{x} - \mathbf{y}\| < \delta'$, then $d(h(\mathbf{x}), h(\mathbf{y})) < \varepsilon$. Let l' denote the maximum of the side-lengths of R'. Recall that, by subdividing R, we have subdivided the whole triangulation in the way described, and in particular we have subdivided R' into 4^m smaller triangles. If now we choose m' with $2^{-m'}l' < \delta'$, we may perform the operation (of subdivision at the midpoints of sides) m' times on each of these smaller triangles, and extending suitably to the rest of the triangulation. In this way, we can replace our original triangulation by one with $4^{m+m'}$ times as many triangles, with the property that both $\hat{R}$ and $\hat{R}'$ have now been subdivided into topological triangles of diameter less than ε. Repeating this process in turn for each of the triangles in the original triangulation, we obtain a triangulation of X, with the same Euler number as our original one, but whose topological triangles all have diameters less than ε. $\qquad\square$

3.3 Polygonal decompositions

Whilst triangulations are the natural decompositions of a compact space to consider from the point of view of Algebraic Topology, we shall also wish to consider decompositions of surfaces into polygons. We shall make the definition here just for S^2 and T, but there will be an obvious generalization for arbitrary compact surfaces as studied in Chapter 8.

Definition 3.12   A *polygonal decomposition* of $X = S^2$ or T consists of a finite collection of polygons which cover X, and whose interiors (the *faces*) are disjoint. The *edges* in the decomposition correspond to sides of the polygons, and the *vertices* in the decomposition to vertices of the polygons. Moreover, we stipulate that the interior of each edge contains no vertices, and is a side of just two polygons in the decomposition.

Our definition implies for instance that the two end-points of an edge must be vertices. The Euler number of the polygonal decomposition is defined precisely as before, namely

$$e = F - E + V,$$

where $F = \#$ faces, $E = \#$ edges and $V = \#$ vertices.

   It will turn out that it is just as valid to calculate Euler numbers of compact surfaces by means of polygonal decompositions rather than triangulations. For the polygonal decomposition of the torus into eight triangles that we gave above (which was not a triangulation), we have $F = 8$, $E = 12$ and $V = 4$, and thus $e = 0$ as expected.

   The method of proof we adopt here to show that the Euler number is (at least for S^2 and T) independent of the choice of triangulation is to replace any given triangulation by a *polygonal decomposition*. This replacement is done in such a way as to ensure that the Euler number of the triangulation is the same as the Euler number of the polygonal decomposition. We then use the following result.

Proposition 3.13   *Suppose we have a polygonal decomposition of S^2, respectively T, and we assume that the Gauss–Bonnet formula holds for the polygons in the decomposition, then the Euler number of the decomposition $e = F - E + V$ is 2, respectively 0.*

Proof   We denote the polygons by $\Pi_1, \ldots, \Pi_F$, where Π_i is assumed to have n_i sides. If the interior angles of Π_i sum to τ_i, then clearly $\sum \tau_i = 2\pi V$, since at each vertex the angles sum to 2π.

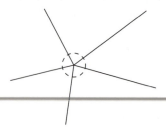

Also note that $\sum_{i=1}^{F} n_i = 2E$ (counting edges), and so $2E - 2F = \sum_i (n_i - 2)$. In the case of S^2, the Gauss–Bonnet formula states that

$$\text{area } \Pi_i = \tau_i - (n_i - 2)\pi.$$

Therefore

$$4\pi = \sum_{i=1}^{F} \text{area } \Pi_i = \sum_i (\tau_i - (n_i - 2)\pi)$$

$$= 2\pi V - \pi(2E - 2F) = 2\pi V - 2\pi E + 2\pi F,$$

and hence $e = 2$.

In the case of the torus T, the Gauss–Bonnet formula states that $\tau_i = (n_i - 2)\pi$ for all i. Therefore

$$2\pi V = \sum_i \tau_i = \sum_i (n_i - 2)\pi = (2E - 2F)\pi,$$

and so $e = 0$.   □

Example   If we are given a convex polyhedron $K \subset \mathbf{R}^3$, for instance one of the five regular solids, we may assume that the origin is in the interior of K and project radially outwards onto the surface of a sphere S^2, also centred on the origin. In this way, we obtain a polygonal decomposition of S^2, whose faces, edges and vertices correspond to the faces, edges and vertices of K. Therefore, we recover the well-known fact that, for any convex polyhedron K, the Euler number $e(K) = 2$.

Suppose now we are given an arbitrary triangulation of $X = S^2$ or T. Our strategy for showing that its Euler number is independent of the triangulation is to reduce to Proposition 3.13, replacing the triangulation by a polygonal decomposition of X with the same Euler number. For that step, we shall need the following result.

Claim   We can replace each edge of a given triangulation by a simple polygonal curve (with the same end-points) in such a way that:

(i)   These polygonal curves have no further points of intersection, and each topological triangle has been replaced by a polygon (whose boundary is just the resulting polygonal approximation to the boundary of the topological triangle).

(ii)   If a triangle of the triangulation is contained in a given open subset of X, then we may ensure (by taking a sufficiently accurate polygonal approximation) that the corresponding polygon is contained in the same open subset.

(iii) The polygons obtained yield a polygonal decomposition of X. An easy check verifies that the Euler number of this decomposition is the same as for the original triangulation.

The general ideas behind the proof of this claim are very intuitive, but care needs to be taken to get the details correct. The construction of the polygonal approximations to the edges of the triangulation is of necessity slightly fiddly, since the edges of the topological triangles in X may be highly non-trivial. A proof of this claim has been placed in an appendix at the end of this chapter. In order to maintain momentum, the reader is recommended to take the claim on trust, safe in the knowledge that there are not any deep ideas involved.

If now we know that the Gauss–Bonnet formula holds for all the polygons in the resulting decomposition, then, by Proposition 3.13, we deduce that the Euler number is 2 for the sphere and 0 for the torus (independent of the triangulation).

We recall however that any open ball on S^2 of radius less than $\pi/2$ is convex in S^2 and that any open ball in T of radius less than $1/4$ is convex in T. It will in fact be true for general compact surfaces, when we come to them, that open balls of small enough radius will be convex. For polygons contained in such convex balls of S^2, respectively T, we have however verified that the Gauss–Bonnet formula holds (Theorem 2.16, respectively Lemma 3.5). This enables us to deduce below (at least for the sphere and torus) that the Euler number is independent of the triangulation. In particular, this implies that the sphere and the torus are not homeomorphic (since any triangulation on a space induces one with the same Euler number on any homeomorphic space).

Theorem 3.14   *Any triangulation of the sphere has Euler number 2. Any triangulation of the torus has Euler number 0.*

Proof Let us take $\varepsilon = \pi/2$ for the sphere, and $\varepsilon = 1/4$ for the torus, and let us denote by X the space being considered (the sphere or torus).

Given any triangulation of X, we may subdivide it so as to obtain a triangulation of X, whose Euler number is unchanged but whose triangles all have diameter less than ε (Proposition 3.11). The above claim then shows that this triangulation may be replaced by a polygonal decomposition of X with the same Euler number, any of whose constituent polygons is contained in some convex open ball of radius ε. Since the Gauss–Bonnet formula holds for such polygons, the theorem follows from Proposition 3.13.   □

Remark   We remark that the proofs in this section and its associated appendix will translate virtually unchanged to the general case of compact surfaces, as studied in Chapter 8, once we have proved the existence of suitable convex open neighbourhoods. The proofs here have deliberately been written in such a form that no significant changes will be required.

Remark   The above theorem is of course a purely topological one, and is independent of any choice of metric on the sphere or torus. The proof we have given does however involve the choice of convenient metrics; this feature, that the proof of a topological result involves a choice of metric, occurs commonly in more advanced geometry. Moreover, in dimensions three and more, the search for 'convenient' metrics is a very active field of current research, as evidenced by the recent solution to the Poincaré conjecture.

3.4   Topology of the g-holed torus

The sphere and torus may be regarded as topological building blocks for other compact surfaces, such as the g-holed tori, for $g \geq 2$. The number g is called the *genus* of the surface. In this section we explain the topology of the g-holed torus, showing that the surface may be obtained by a standard gluing construction. In Chapter 8, we shall extend these ideas to include the gluing of these building blocks equipped with convenient metrics, thereby also gluing the metrics and obtaining a geometric rather than just topological understanding of the surfaces.

For simplicity, we describe explicitly the case when $g = 2$; the higher genus cases will represent an easy extension of this case. Let us take the 2-holed torus, embedded in $\mathbf{R}^3$, as illustrated in Section 3.1. If we cut along the dotted curve, we get the two surfaces as illustrated below, where each of the two surfaces is obtained topologically from a torus by removing a disc. Thus the 2-holed torus is obtained from these two punctured tori, by identifying them along their boundaries (i.e. by gluing the two circles back together).

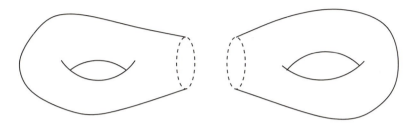

If we are only interested in this gluing from the topological point of view, we can triangulate the two tori, remove a triangle from both, and then identify along the common boundary. Here, it is more natural to regard the result as a topological space rather than a metric space, as gluing constructions usually distort distances. It is then clear that the triangulations on the two (punctured) tori match up to give a triangulation on the 2-holed torus. The triangulations on the disjoint union of the (unpunctured) tori yield an Euler number of 0; we are removing two faces, and making the identification on the triangles, then reducing the number of edges and vertices both by three. The Euler number of this triangulation is therefore -2; in Chapter 8, we shall see that the Euler number is independent of the triangulation, and is therefore always -2.

Let us now represent the two torus by squares, with sides appropriately identified. We may assume that we remove from each torus a topological disc, whose boundary passes through the point on the torus represented by the corners of the square. This is illustrated by the diagram below, where the regions bounded by the dotted curves and all the vertices of the squares have been removed.

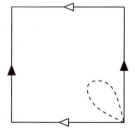 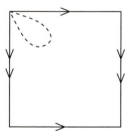

We may open out both the dotted curves to produce two pentagons; if we now reinstate all the missing points on the boundaries of these two pentagons, we are in fact reinstating the boundary S^1 for each punctured torus. Gluing these boundary circles together now corresponds to gluing the two pentagons together along the dotted sides, this gluing being represented in the diagram below by the curved double headed arrow, to yield an octagon with sides identified as shown.

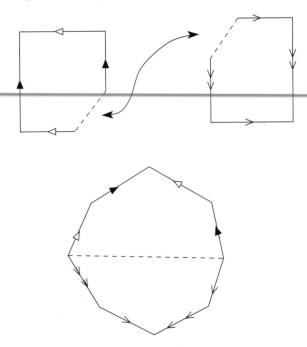

All this may now be extended to the case of genus three or more. Thus a g-holed torus is obtained by gluing, in an obvious way, two singly punctured tori to $g - 2$ doubly punctured tori. Representing the punctures by (disjoint) triangles in a triangulation on each of the tori, we observe that a triangulation of the g-holed torus exists, with Euler number $2 - 2g$. An inductive argument (just extending the argument we gave above) shows that topologically the g-holed torus may be obtained from a $4g$-gon by grouping the sides into g sets of four adjacent sides, and within each of these sets of four adjacent sides making identifications in an analogous way to the cases when $g = 1$ and 2.

All the vertices of this $4g$-gon are identified to a single point on the surface of genus g. In the case of $g = 1$, the angles of the Euclidean square sum to 2π. It was for this reason that we could have a locally Euclidean metric induced on the torus from the Euclidean metric on the square. A small open ball round the point of the torus represented by the vertices of the square will then be isometric to a small Euclidean ball.

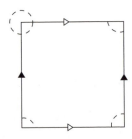

This explains also why it is not possible to induce, by an analogous construction, a locally Euclidean metric on a surface of genus $g \geq 2$, as the angles at the vertices of the Euclidean $4g$-gon sum to more than 2π. We have topologically identified the surface as a plane $4g$-gon with sides identified, but these identifications must involve distorting distances. In fact, the general Gauss–Bonnet theorem, as proved in Chapter 8, provides an explicit obstruction to locally Euclidean metrics being defined on the surface, given that a surface of genus $g \geq 2$ has negative Euler number. We shall observe later that it is possible however to define a *locally hyperbolic* metric on such a surface (see comment following Proposition 5.23).

One of the standard results proved in any first course on Algebraic Topology is the classification of compact topological surfaces up to homeomorphism — see for instance [7], Chapter I, Section 7. This classification yields two series of surfaces. The *orientable* surfaces are just the g-holed tori (for $g \geq 0$) described above. There are however the *non-orientable* examples, which are obtained by removing g disjoint open discs from S^2 and gluing in g copies of the Möbius strip. We recall that the Möbius strip is defined by identifying a pair of opposite sides of a square, but in an opposite way to that which yields a cylinder, and that its boundary is topologically just a copy of S^1. The first, and most geometric, of these compact non-orientable examples is the real projective plane.

Example (The real projective plane) The *real projective plane* $\mathbf{P}^2(\mathbf{R})$ has points corresponding to lines through the origin in $\mathbf{R}^3$. Equivalently, it may be defined as $S^2 / \sim$, where $\sim$ denotes the equivalence relation which identifies antipodal points. We can then define a metric on $\mathbf{P}^2(\mathbf{R})$, coming from the spherical metric on S^2; given points of $\mathbf{P}^2(\mathbf{R})$ coming from points $\mathbf{x}, \mathbf{y}$ in S^2, the distance between them is just

$$\min\{d(\pm\mathbf{x}, \pm\mathbf{y})\},$$

where d here represents the spherical metric on S^2. The key observation is that locally this is precisely the same as the spherical metric, since any ball of radius less than $\pi/4$ in $\mathbf{P}^2(\mathbf{R})$ is isometric to a corresponding ball on S^2. The quotient map $S^2 \to S^2/\sim$ is then a continuous surjection, and thus $\mathbf{P}^2(\mathbf{R})$ is compact. The quotient map is a 2-1 local homeomorphism, and we say that the real projective plane has S^2 as a *double cover*.

An equivalent representation of the real projective plane is as the closed northern hemisphere on S^2, but with antipodal points on the equator identified. On this model of the projective plane, the metric is essentially given by the spherical metric, but one is given 'free transport' between antipodal points on the equator.

In some ways, the geometry of $\mathbf{P}^2(\mathbf{R})$ is nicer than that of S^2. A *geodesic line* in $\mathbf{P}^2(\mathbf{R})$ corresponds by definition to a plane through the origin in $\mathbf{R}^3$, or equivalently to a great circle on S^2. As was the case for S^2, two lines always meet, but for $\mathbf{P}^2(\mathbf{R})$ they meet in exactly one point (since antipodal points on S^2 represent the same point of $\mathbf{P}^2(\mathbf{R})$).

With the interpretation of the projective plane in terms of the northern hemisphere model, it is easy to write down a (geodesic) triangulation of $\mathbf{P}^2(\mathbf{R})$, by for instance subdividing the hemisphere into eight segments, where antipodal points and edges are identified. The diagram below shows the view of this from above, with the central point representing the north pole.

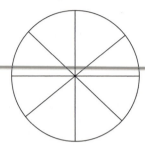

There are 8 triangles in this triangulation, and (taking into account identifications) there are 12 edges and 5 vertices, and hence an Euler number of 1. In fact, using Constructions 3.9 and 3.15 and arguing as in the case of the sphere, we see that, for any topological triangulation of the real projective plane, we can firstly subdivide it so that every topological triangle is contained in some convex ball of radius less than $\pi/4$, and then approximate each triangle by a polygon (also contained in the convex ball). Since these are now spherical polygons, we have the Gauss–Bonnet formula for their areas, and the argument from Proposition 3.13 shows that the Euler number is one (corresponding to the fact that the area of $\mathbf{P}^2(\mathbf{R})$ is just the area of the hemisphere, namely 2π).

The real projective plane may also be represented topologically by a square with opposite sides identified as below (corresponding to the identification of antipodal points on the boundary).

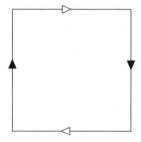

This may be compared with the identifications of the sides which produced the torus; in this case however, only diagonally opposite corners are identified, as opposed to all four corners for the torus. Also, unlike the identifications for the torus, we cannot make the identifications to get the projective plane without distorting distances on the square.

There are two other ways of identifying the sides of the square, other than those leading to the torus and the real projective plane. One of these just gives the sphere, and the remaining one is the *Klein bottle*, as illustrated below.

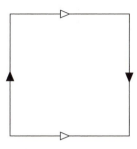

If we identify the top and bottom of the square as shown, we obtain a cylinder with circular ends. We now identify these circular ends, but in the opposite way to that which we used to get the torus. It will still be the case however that the vertices of the square are all identified to a single point on the Klein bottle, and that we can define a locally Euclidean metric on the Klein bottle (Exercise 3.11). The Euler number of the Klein bottle may be checked to be zero (Exercise 3.10).

Exercises

3.1 By considering the circumferences of small circles, show that the sphere and the (locally Euclidean) torus do not contain non-empty open subsets which are isometric to each other.

3.2 For any two distinct points on the locally Euclidean torus T, show that there are infinitely many geodesic lines joining them.

3.3 Suppose Q_1 is a closed square in $\mathbf{R}^2$ with vertices (p, q), $(p + 1/2, q)$, $(p, q + 1/2)$, $(p + 1/2, q + 1/2)$. With T identified as the quotient of $\mathbf{R}^2$ by a unit square lattice, show that the restriction to Int Q_1 of the quotient map $\varphi : \mathbf{R}^2 \to T$ is an *isometry* of Int Q_1 onto its image. Show that the image of Int Q_1 is a convex open subset of T.

3.4 Let T denote the locally Euclidean torus defined by the unit square lattice in $\mathbf{R}^2$. Show that T has the structure of an abelian group, and that it may be identified as a subgroup of its isometry group Isom(T); deduce that Isom(T) acts transitively on T. Show further that the group of isometries fixing a given point of T is a dihedral group of order eight, i.e. the full symmetry group of the square.

3.5 Using radial projection from the centre of S^2, or otherwise, show that any spherical triangle is a topological triangle on S^2.

3.6 Let T denote the locally Euclidean torus defined by the unit square lattice in $\mathbf{R}^2$. Given integral vectors $\mathbf{m} = (m_1, m_2)$ and $\mathbf{n} = (n_1, n_2)$ with $m_1 n_2 - m_2 n_1 = 1$, and an arbitrary vector $\mathbf{a} \in \mathbf{R}^2$, let Π be the parallelogram with vertices at $\mathbf{a}, \mathbf{a} + \mathbf{m}, \mathbf{a} + \mathbf{n}$ and $\mathbf{a} + \mathbf{m} + \mathbf{n}$. Show that the quotient map $\varphi : \mathbf{R}^2 \to T$ restricts to a homeomorphism of Int Π onto its image. Deduce the existence of convex polygons on T which are not contained in the image under φ of any unit square.

3.7 Suppose we have a polygonal decomposition of S^2 or T. We denote by F_n the number of faces with precisely n edges, and V_m the number of vertices where precisely m edges meet. If E denotes the total number of edges, show that $\sum_n nF_n = 2E = \sum_m mV_m$.

We suppose that each face has at least three edges, and at least three edges meet at each vertex. If $V_3 = 0$, deduce that $E \geq 2V$, where V is the total number of vertices. If $F_3 = 0$, deduce that $E \geq 2F$, where F is the total number of faces. For the sphere, deduce that $V_3 + F_3 > 0$. For the torus, exhibit a polygonal decomposition with $V_3 = 0 = F_3$.

3.8   With the notation as in the previous exercise, given a polygonal decomposition of S^2, prove the identity

$$\sum_n (6 - n)F_n = 12 + 2\sum_m (m - 3)V_m.$$

If each face has at least three edges, and at least three edges meet at each vertex, deduce the inequality $3F_3 + 2F_4 + F_5 \geq 12$.

The surface of a football is decomposed into spherical hexagons and pentagons, with precisely three faces meeting at each vertex. How many pentagons are there? Demonstrate the existence of such a decomposition with each vertex contained in precisely one pentagon.

3.9   Find an example of two distinct circles of radii less than $\pi/2$ in the real projective plane which meet in four points.

3.10   Find a triangulation for the Klein bottle, and check that its Euler number is zero.

3.11   Prove that there exists a continuous surjective map from the torus to the Klein bottle, which is a 2-1 local homeomorphism (i.e. the Klein bottle has the torus as a double cover). Hence, or otherwise, show that a locally Euclidean metric may be defined on the Klein bottle.

Appendix on polygonal approximations

In this appendix, which is included for the sake of completeness, we give a full proof of the claim from Section 3.3. This allowed us to replace a triangulation on our surface X, the sphere or torus equipped with the given metrics, by an associated polygonal decomposition with the same Euler number. In Construction 3.15, we explain in detail how to polygonally approximate the edges of the triangulation, and in Proposition 3.16 we show that the construction yields a polygonal decomposition of X with the properties claimed. In the construction, it is crucial that one first approximates, by geodesic line segments, those parts of the edges which are near the vertices. After this has been done, finding a good approximation to the remaining parts of the edges (away from the vertices) is reasonably straightforward. If say the edges were smooth curves near the vertices with distinct tangent directions, the first part would also be straightforward. In general however, the edges may be given by rather complicated curves which are only continuous, and so a slightly more subtle argument is needed. In the proof of Proposition 3.16, we use winding numbers (as introduced in Chapter 1) to identify the interior of the polygons, and to relate the polygons to the topological triangles in the triangulation. As remarked before, the material in this appendix will translate virtually unchanged to the general case of compact surfaces in Chapter 8, once we have the existence of suitable convex open neighbourhoods.

Construction 3.15   Let us denote the topological triangles in the triangulation by $\hat{R}_j = f_j(R_j)$, where $R_j \subset \mathbf{R}^2$ is a Euclidean triangle and U_j is some ϵ_j-neighbourhood of R_j, and $f_j : U_j \to V_j \subset X$ is a homeomorphism of U_j onto an open subset of X (for the present, X being the sphere or torus). The index j will range from 1 to the number of faces F. For each j, we shall fix a reference point z_j in the interior of R_j — the barycentre (i.e. centroid) of R_j will be a convenient choice for such a point — and let d_j be the distance of z_j from the boundary of R_j. Let $\hat{z}_j = f_j(z_j)$ denote its image in $\hat{R}_j$. We may assume also that U_j has been chosen to be a sufficiently small neighbourhood of R_j so that the open set $V_j = f_j(U_j)$ in X contains none of the points $\hat{z}_k$ for $k \neq j$, nor any of the vertices of the triangulation other than the three vertices of $\hat{R}_j$. If there were such points $\hat{z}$ in V_j, then we just need to choose ϵ_j sufficiently small such that the ϵ_j-neighbourhood of R_j avoids all the pre-images in U_j (true if the ϵ_j-balls round these pre-images are disjoint from R_j).

The crucial idea of this construction is that we should first modify the edges of the triangulation *near the vertices*. Let us pick therefore a vertex P and consider all the topological triangles $\hat{R}_j$ with P as a vertex, which by reordering we assume are the $\hat{R}_j$ with $j = 1, \ldots, s$. For each $1 \leq j \leq s$, we choose a small disc $D_j \subset U_j$ around the vertex of R_j corresponding to P (with the radius of D_j being also assumed to be less than $d_j/2$). Each $f_j(D_j)$, for $j = 1, \ldots, s$, is an open neighbourhood of P in X, and so we may choose a convex open ball $B(P, \delta)$ around P contained in them all (with δ here depending on the vertex P). We may label the edges in the triangulation with P as an end-point by $C_1, \ldots, C_s$, where C_i and C_{i+1} are edges of the topological triangle $\hat{R}_i$ (and with C_s and C_1 being edges of $\hat{R}_s$). We assume also that δ has been chosen sufficiently small so that these are the only edges of the triangulation intersecting $B(P, \delta)$.

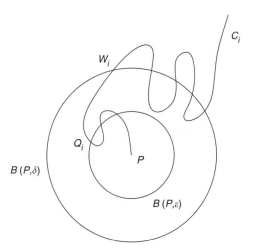

For $1 \leq i \leq s$, we consider the *first* point W_i where C_i meets the boundary of $\bar{B}(P, \delta)$, as one travels from P along C_i. We can now choose $0 < \varepsilon < \delta$ such that the distance $\rho(P, x) > \varepsilon$ for all points x of C_i beyond W_i — here, we are using the strict positivity of the distance of P from a closed set not containing P (namely that part of C_i consisting of W_i and the points beyond W_i). Moreover, we choose $\varepsilon > 0$ so that the above properties hold for all $i = 1, \ldots, s$, where $\varepsilon = \varepsilon(P)$ depends only on the vertex P.

We now let Q_i be the *last* point (coming from P) where C_i meets the boundary of $\bar{B}(P, \varepsilon)$, or alternatively the first such point encountered when travelling in the opposite direction (from W_i). We deduce that all points on C_i beyond Q_i (travelling from P) lie outside $\bar{B}(P, \varepsilon)$, and that the part of C_i up to Q_i is contained in $B(P, \delta)$. We now *replace* the section of C_i between P and Q_i by the geodesic line segment PQ_i. We repeat the above procedure at each of the vertices of the triangulation.

The edges C_j in the triangulation have now been replaced by approximations γ_j, which are disjoint except for the vertices at their end-points. We now modify these curves in turn, replacing them by polygonal approximations. Consider one such curve $\gamma = \gamma_i$, given by a continuous map $\gamma : [0, 1] \rightarrow X$. By construction, $\gamma(0)$ and $\gamma(1)$ were vertices of the triangulation, and there exist $\kappa_1, \kappa_2 > 0$ such that on both $[0, \kappa_1]$ and $[1 - \kappa_2, 1]$, the curve γ is a geodesic line segment. Moreover, the curve γ replaced an edge C_i of the triangulation, with C_i being an edge of two of the topological triangles, say $\hat{R}_k$ and $\hat{R}_l$. The curve segment $G = \gamma([\kappa_1, 1 - \kappa_2])$ is contained in the complement of all the other curves γ_j for $j \neq i$ (an open set). For each point y of this intermediate segment of γ, we can choose a convex open ball B around y with the following properties:

- B is disjoint from the other curves γ_j for $j \neq i$.
- B is contained in both $V_k = f_k(U_k)$ and $V_l = f_l(U_l)$, where the indices k and l refer to the two triangles specified above.
- B is disjoint from the closed balls $\bar{B}(P, \varepsilon(P)/2)$ defined above, for all vertices P of the triangulation.
- For both $j = k$ and l, the inverse image $f_j^{-1}(B)$ is contained in a ball in U_j around the boundary point $f_j^{-1}(y)$ on R_j of radius less than $d_j/2$.

Moreover, since the curve segment $G = \gamma([\kappa_1, 1 - \kappa_2])$ is compact by Lemma 1.15, it may be covered by a finite number of such convex open balls, $B_1, \ldots, B_m$ say. Assuming that no proper subcollection of the B_r cover G, we may order the B_r so that

$$G \cap (B_1 \cup \ldots \cup B_r) = \gamma([\kappa_1, s_r)) \qquad \text{for } r = 1, \ldots, m - 1,$$
$$= G \qquad \text{for } r = m,$$

where $\kappa_1 < s_1 < s_2 < \ldots < s_{m-1} < 1 - \kappa_2$. We can then find a dissection

$$\kappa_1 = t_0 < t_1 < \ldots < t_{m-1} < t_m = 1 - \kappa_2$$

of this closed interval so that $\gamma(t_r) \in B_r \cap B_{r+1}$ for all $1 < r < m$ (for each $r \leq m - 1$, we have $s_{r-1} < t_r < s_r$). Joining each $\gamma(t_r)$ to $\gamma(t_{r+1})$ by a geodesic line segment in B_r, we may replace the curve segment in question by a polygonal approximation, and hence the whole edge by a polygonal approximation $\tilde{\gamma} : [0, 1] \rightarrow X$, where the image of the open interval $(0, 1)$ is disjoint from the other curves γ_j for $j \neq i$. It might happen however that the polygonal curve $\tilde{\gamma}$ constructed in this way is no longer simple, but this is not a problem: if $\tilde{\gamma}(\sigma_1) = \tilde{\gamma}(\sigma_2)$ for some $\sigma_1 < \sigma_2$, we may just omit the part of $\tilde{\gamma}$ between σ_1 and σ_2. In this way, we have replaced γ by a *simple* polygonal curve γ^*. Observe that if the vertex P is an initial point for γ^*, then the only part of γ^* which lies in $B(P, \varepsilon(P)/2)$ is an initial geodesic line segment.

We apply this procedure in turn to all the curves γ_j, with the following convention. When choosing our convex open balls B as above, we shall want B to be disjoint from the curves γ_i^* which have already been polygonally approximated, as well as from the curves γ_i which are yet to be approximated. On completion of this step, we have replaced all the edges in our original triangulation by polygonal approximations. For a given topological triangle $\hat{R}_j$, the boundary curve Γ_j (with initial and final point at some vertex) has been replaced by a simple closed polygonal approximation Γ_j^*, formed by three simple polygonal curves of the form γ_i^*.

The central claim we made in Section 3.3 then follows from the following result.

Proposition 3.16   *With the notation as in the above construction, each Γ_j^* is the boundary of a unique polygon on X with $\hat{z}_j = f_j(z_j)$ in its interior, and not containing any of the other reference points $\hat{z}_k$ for $k \neq j$. These polygons form a polygonal decomposition of X. Moreover, if a topological triangle of the triangulation is contained in a given open subset of X, then we may ensure that the corresponding polygon is contained in the same open subset.*

Proof   For a given j, the boundary of the Euclidean triangle R_j corresponds under f_j to the curve Γ_j, and we denote by Υ_j the continuous closed curve in U_j corresponding to the polygonal approximation Γ_j^*. We observe that Υ_j consists of three segments η_1, η_2 and η_3, corresponding to simple polygonal curves γ_k^* as constructed above. Each η_i may be regarded as replacing a side L_i of the triangle $R_j \subset U_j$.

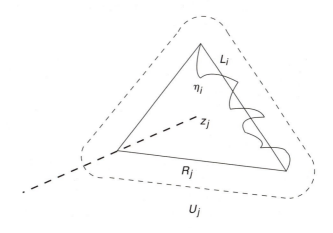

By construction of the polygonal approximation (in particular, the choice of the convex open balls B that was made), every point of η_i is within a distance $d_j/2$ of the side L_i. Thus the closed curve in U_j given by concatenating η_i with the line segment L_i taken in the opposite direction is contained in the complement in $\mathbf{C} = \mathbf{R}^2$ of the semi-infinite ray starting at z_j and passing though the vertex of R_j opposite the side L_i. Elementary properties of the winding number imply that this closed curve has winding number zero about z_j (since, if we take z_j as the origin, there is a continuous branch of the argument on the complement of this ray).

The boundary of the triangle R_j has winding number ± 1 about z_j. We may however replace each side L_i in turn by the corresponding curve η_i without changing the winding number; here we are using the property, stated in Chapter 1, concerning the winding number of the concatenation of two curves. Applying this to all three sides L_i, we see that the winding number of Υ_j about z_j is still ± 1.

The argument of Proposition 1.17 shows that the complement of the simple closed polygonal curve Γ_j^* in X has at most two connected components; if there are two, these correspond to the two sides of Γ_j^*. Since the simple closed curve Υ_j has winding number ± 1 about z_j, it follows that the complement of Υ_j in $\mathbf{C}$ has two components, the bounded one of which contains z_j. Since the complement of U_j in $\mathbf{C}$ is connected by Remark 3.7 (recall that we took U_j to be some ϵ_j-neighbourhood of R_j), it is contained in the unbounded component; in particular, we deduce that the bounded component of the complement of Υ_j in $\mathbf{C}$ is contained in U_j, and therefore specifies a polygon on X, with $\hat{z}_j$ in its interior. Moreover, this polygon contains none of the other points $\hat{z}_k$ for $k \neq j$, nor any of the vertices of the original triangulation other than the three vertices of $\hat{R}_j$, since any such point would be in U_j, contrary to our initial assumptions.

We now need to show that these polygons yield a polygonal decomposition of X. We *claim* first that if Π_1 and Π_2 are two of the polygons constructed above, corresponding say to topological triangles $\hat{R}_1$ and $\hat{R}_2$ of our original triangulation, then *the interior of* Π_1 *contains no points of the boundary of* Π_2. If $\hat{R}_1$ and $\hat{R}_2$ have an edge C in common, then the boundaries of Π_1 and Π_2 have a corresponding simple polygonal curve γ^* in common, and this is disjoint from the interior of Π_1. Otherwise, an edge C of $\hat{R}_2$ gives rise to a simple polygonal curve γ^* (part of the boundary of Π_2) with at least one end-point not contained in Π_1 (since, from the defining properties of triangulations, it is a vertex of the original triangulation other than the three vertices of $\hat{R}_1$, and hence by construction is not contained in Π_1). We deduce that γ^* cannot contain points in the interior of Π_1; if so, the curve γ^* would intersect the boundary of Π_1 for some intermediate point. However, by construction, a point in the interior of γ^* cannot be a point on the boundary of Π_1 (otherwise two of the curves γ_j^* would intersect at a point other than a common end-point).

This now implies that the interiors Int Π_1 and Int Π_2 are *disjoint*. If there was a point $\hat{z}$ in the intersection, then we could find a curve ξ joining $\hat{z}_1$ to $\hat{z}$ in Int Π_1. Since by construction $\hat{z}_1 \notin$ Int Π_2, the curve ξ must intersect the boundary of Π_2 at some point in the interior of Π_1, contradicting what we proved in the previous paragraph.

Each *edge* χ of the putative polygonal decomposition is a segment of a simple polygonal curve γ^*, approximating an edge C of the original triangulation, an edge for precisely two of the topological triangles. Thus χ is certainly a side for the two resulting polygons. If it were a side for more than two of the polygons, then there would be polygons with non-disjoint interiors, contrary to what we have just proved.

To see that a polygonal decomposition of X has been obtained, it remains to show that X is the union Y of the closed polygons $\Pi_1, \ldots, \Pi_F$. Clearly Y is a closed set; we show that Y is also open, and hence by connectedness that $Y = X$. If a point $P \in Y$ is in the interior of a polygon, then clearly some open neighbourhood is contained in Y. If $P \in Y$ is not a vertex of our original triangulation but not in the interior of any polygon of the decomposition, then the above reasoning shows that it is an interior point of one of the simple polygonal curves γ^* and on the boundary of precisely two of the polygons; from this, it follows again that some open neighbourhood of P is contained in Y. So we need to consider the case when P is a vertex of the original triangulation; if $\Pi_1, \ldots, \Pi_s$ are the polygons with P as a vertex, Construction 3.15 ensures that the open ball $B(P, \varepsilon(P)/2)$ is the union of sectors $B(P, \varepsilon(P)/2) \cap \Pi_i$, for $i = 1, \ldots, s$, and hence is contained in Y. Thus Y is open as claimed, and we have a polygonal decomposition of X.

The final sentence of the proposition is easy. A given topological triangle $\hat{R}$ of the triangulation is the image of a Euclidean triangle R under a homeomorphism $f : U \to V$, from some ϵ-neighbourhood U of R in $\mathbf{R}^2$ to an open subset V of X. If however $\hat{R}$ is contained in an open set V' of X, then R is contained in an open subset $f^{-1}(V')$ of U. Using the fact (see Exercise 1.16) that the compact set R has strictly positive distance from the complement of $f^{-1}(V')$ in $\mathbf{R}^2$ (a closed set), we may find an ϵ'-neighbourhood of R contained in $f^{-1}(V')$. Thus, we may ensure, by taking our original ϵ sufficiently small, that the ϵ-neighbourhood U has image $V \subset V'$. Since the polygon we construct corresponding to $\hat{R}$ is contained in V, we have ensured that it is contained in our given open subset V'. $\qquad\square$

4 Riemannian metrics

A central concept in this book is that of the length of a curve. So far, we have only needed lengths explicitly for rather simple curves. For more general curves, even in $\mathbf{R}^2$, we shall need to invoke Proposition 1.10, in which the length of a smooth curve Γ was interpreted as an integral of $\|\Gamma'(t)\|$; we therefore need calculus to do the integration.

Another reason for needing calculus is that in general the geometry of the space may also vary from point to point. If one thinks about a map of an area of Britain, it is not entirely accurate in predicting lengths of journeys, just from the scale and measuring distances on the map; in other words, the distance on the ground is not just a scaled up version of the Euclidean distance on the map. One reason for this is because of the curvature of the earth, but for maps of the scale being discussed, this should not be too significant. Another reason is however the nature of the terrain being mapped. Around the city of Cambridge in the East of England, the land is very flat, and so the map provides a fairly good measure of the lengths of journeys. If however the map were of part of North Wales, a terrain with many mountains and valleys, it would be very inaccurate to measure the lengths of journeys just from the length of the corresponding curve on the map. The map alleviates this by providing contour lines, and a skilled map-reader can use these to estimate the length of a journey. What we need however for accurate calculations is information about the scaling required in each direction at each point. This leads to the idea that when taking the norm of $\Gamma'(t) \in \mathbf{R}^2$, this norm should be dependent on the point $\Gamma(t)$ on the map. This idea of a smooth family of norms on $\mathbf{R}^2$, the norms depending on points of some open subset of the plane, leads to the concept of a *Riemannian metric*, which will be vital for the rest of the book.

The ideas introduced in this chapter will be immediately reinforced in Chapter 5, where we make a detailed study of an important example, namely the *hyperbolic plane*.

4.1 Revision on derivatives and the Chain Rule

Before formalizing the concept of a Riemannian metric, we should recall various facts from Analysis about differentiating functions in several variables. This section provides such revision as might be needed, but also allows us to fix on the most convenient notation for derivatives, at least for our purposes. The right choice of notation enables us for instance to give a precise meaning to the idea of *differentials*, a concept treated rather cavalierly in some undergraduate textbooks, especially in Applied Mathematics.

Suppose U is an open subset of $\mathbf{R}^n$; a map $f : U \to \mathbf{R}^m$ is defined coordinatewise by real-valued functions $(f_1, \ldots, f_m)$ on U. The map f is called *smooth* (or C^∞) if each f_i has partial derivatives of all orders. A smooth map is certainly differentiable (for which having continuous partial derivatives suffices). The *derivative* of f at $\mathbf{a} \in U$ is a linear map $df_\mathbf{a} : \mathbf{R}^n \to \mathbf{R}^m$ (in some books, denoted $Df_\mathbf{a}$ or $f'(\mathbf{a})$) such that, for $\mathbf{h} \neq \mathbf{0}$,

$$\frac{\|f(\mathbf{a} + \mathbf{h}) - f(\mathbf{a}) - df_\mathbf{a}(\mathbf{h})\|}{\|\mathbf{h}\|} \to 0 \quad \text{as} \quad \mathbf{h} \to \mathbf{0} \in \mathbf{R}^n.$$

When $m = 1$, the linear map $df_\mathbf{a} : \mathbf{R}^n \to \mathbf{R}$ is determined by the partial derivatives of f at $\mathbf{a}$, namely $\left(\frac{\partial f}{\partial x_1}(\mathbf{a}), \ldots, \frac{\partial f}{\partial x_n}(\mathbf{a}) \right)$ via matrix multiplication, i.e.

$$(h_1, \ldots, h_n) \mapsto \sum_i \frac{\partial f}{\partial x_i}(\mathbf{a}) \, h_i.$$

In general, when m is arbitrary, $df_\mathbf{a} : \mathbf{R}^n \to \mathbf{R}^m$ is determined by the $m \times n$ matrix of partial derivatives at $\mathbf{a}$, the *Jacobian* matrix

$$J(f) = \left(\frac{\partial f_i}{\partial x_j} \right).$$

Example   We consider analytic functions $f : U \to \mathbf{C}$ in one complex variable z, where U is an open subset of $\mathbf{C}$. By definition, this means that, for any $z \in U$,

$$\frac{|f(z + w) - f(z) - wf'(z)|}{|w|} \to 0 \quad \text{as} \quad 0 \neq w \to 0 \in \mathbf{C},$$

where f' denotes the (complex valued) derivative df/dz. So if, for $P \in U$, we set $f'(P) = a + ib$, and $w = h_1 + ih_2$, then

$$wf'(P) = (ah_1 - bh_2) + i(bh_1 + ah_2).$$

If we now consider f as a map $U \to \mathbf{R}^2$, then the linear map $df_P : \mathbf{R}^2 \to \mathbf{R}^2$ is represented by the matrix

$$\begin{pmatrix} a & -b \\ b & a \end{pmatrix}.$$

Given any smooth real-valued functions $u(x, y)$, $v(x, y)$ on an open set $U \subset \mathbf{R}^2$, and writing $f(x + iy) = u(x, y) + iv(x, y)$, recall that f is an analytic function of

$z = x + iy$ if and only if the *Cauchy–Riemann* equations are satisfied, namely that $\partial u/\partial x = \partial v/\partial y$ and $\partial u/\partial y = -\partial v/\partial x$ hold everywhere. This essentially is the content of the above calculation.

In particular, we note that if f is a complex analytic function on $U \subset \mathbf{C}$, and P a point of U with $f'(P)$ non-zero, then df_P preserves angles and orientations, since all non-zero vectors are rotated through the same angle, namely the argument of the complex number $f'(P)$. If two smooth curves γ_1, γ_2 pass though the point $P \in U$, the angle at which they meet is defined to be the angle between the derivatives γ_1' and γ_2' at P. If the two curves meet at P with an angle α, then use of the Chain Rule shows that their images $f \circ \gamma_1$ and $f \circ \gamma_2$ meet at $f(P)$ with the same angle and the same orientation.

Playing a central role in the rest of the book will be the Chain Rule. Suppose $U \subset \mathbf{R}^n$ and $V \subset \mathbf{R}^p$ are open subsets. Given smooth maps $f : U \to \mathbf{R}^m$ and $g : V \to U$, then the composite $fg : V \to \mathbf{R}^m$ is a smooth map, and has derivative at $P \in V$

$$d(fg)_P = df_{g(P)} \circ dg_P.$$

In other words, this is summed up by the slogan that the derivative of the composite is just the composite of the derivatives. In terms of Jacobian matrices, it says

$$J(fg)_P = J(f)_{g(P)} J(g)_P,$$

where the multiplication here is just matrix multiplication.

Of particular importance will be the case when U is an open subset of $\mathbf{R}^n$ and $f : U \to \mathbf{R}$ is smooth. For each $P \in U$, we have the linear map $df_P : \mathbf{R}^n \to \mathbf{R}$ (an element therefore of the *dual* space to $\mathbf{R}^n$). These then yield a smooth map

$$df : U \to \operatorname{Hom}(\mathbf{R}^n, \mathbf{R}),$$

where $\operatorname{Hom}(\mathbf{R}^n, \mathbf{R})$ denotes the *dual* vector space, consisting of homogeneous linear forms on $\mathbf{R}^n$, where the dual space may also be identified with $\mathbf{R}^n$. More generally, any smoothly varying family of homogeneous linear forms on $\mathbf{R}^n$, parametrized by U (i.e. given by a smooth map $g : U \to \operatorname{Hom}(\mathbf{R}^n, \mathbf{R})$), is called a *differential* on U.

To understand this concept concretely, we should work in coordinates; if the standard basis of $\mathbf{R}^n$ is denoted by $e_1, \dots, e_n$, there are corresponding coordinate functions $x_i : \mathbf{R}^n \to \mathbf{R}$, defined by projection onto the ith coordinate. As these are already homogeneous linear forms on $\mathbf{R}^n$, the derivative $(dx_i)_P$ is the same linear form, independent of P. Thus the differential dx_i is a constant function of P, and may therefore be regarded as a fixed linear form, namely the corresponding coordinate function on $\mathbf{R}^n$, given by

$$dx_i(a_1, \dots, a_n) = a_i.$$

The homogeneous linear forms $dx_1, \ldots, dx_n$ provide the dual basis to the standard basis $e_1, \ldots, e_n$ of $\mathbf{R}^n$, that is $dx_i(e_j) = \delta_{ij}$ for all $1 \leq i, j \leq n$. We may wish to change the origin in $\mathbf{R}^n$, and then the functions x_i may acquire constant terms; however, the corresponding linear forms dx_i are still just the elements of the standard dual basis. A general differential on U may therefore, by definition, be written in the form $\sum_{i=1}^{n} g_i \, dx_i$, with the g_i smooth functions on U, or equivalently as the row vector $(g_1, \ldots, g_n)$ of smooth functions.

If now $f : U \to \mathbf{R}$ is an arbitrary smooth function on an open set $U \subset \mathbf{R}^n$, the linear form $df_P : \mathbf{R}^n \to \mathbf{R}$ is represented by the partial derivatives at P,

$$\left(\frac{\partial f}{\partial x_1}(P), \ldots, \frac{\partial f}{\partial x_n}(P) \right),$$

or equivalently

$$df_P = \sum_i \frac{\partial f}{\partial x_i}(P) \, dx_i.$$

Thus, as differentials, we have the familar identity that

$$df = \sum_i (\partial f / \partial x_i) \, dx_i.$$

The Chain Rule also translates into a familiar identity on differentials. If $g = (g_1, \ldots, g_n) : U \to \mathbf{R}^n$, where U is an open subset of $\mathbf{R}^m$ (coordinates $u_1, \ldots, u_m$ say) and the g_i are smooth functions on U, then for any coordinate function x_i on $\mathbf{R}^n$, we have

$$dx_i = \sum_{j=1}^{m} (\partial g_i / \partial u_j) \, du_j,$$

as differentials on U. We shall use this in particular when $m = n$, and g represents the map corresponding to a change of coordinates $x_i = g_i(u_1, \ldots, u_n)$. We often then write

$$dx_i = \sum_{j=1}^{n} (\partial x_i / \partial u_j) \, du_j.$$

The *moral* therefore of the last few paragraphs is that one may continue manipulating differentials in the familiar formal way, but we now have a rigorous interpretation as to what a differential is: namely, it is a smooth family of (homogeneous) linear forms.

There is a notational convention which we shall use frequently below. If $\alpha : \mathbf{R}^n \to \mathbf{R}$ and $\beta : \mathbf{R}^n \to \mathbf{R}$ are two linear (homogeneous) forms on $\mathbf{R}^n$, we have an obvious quadratic form $\alpha\beta$ given by multiplying the forms. The associated bilinear form will also be denoted $\alpha\beta$, and is given by

$$(\alpha\beta)(\mathbf{x}, \mathbf{y}) := (\alpha(\mathbf{x})\beta(\mathbf{y}) + \alpha(\mathbf{y})\beta(\mathbf{x}))\,/2.$$

For real numbers $\lambda_1, \lambda_2, \mu_1, \mu_2$ and linear forms $\alpha_1, \alpha_2, \beta_1, \beta_2$, we note the equality of bilinear forms

$$(\lambda_1\alpha_1 + \lambda_2\alpha_2)(\mu_1\beta_1 + \mu_2\beta_2) = \lambda_1\mu_1\alpha_1\beta_1 + \lambda_1\mu_2\alpha_1\beta_2 + \lambda_2\mu_1\alpha_2\beta_1 + \lambda_2\mu_2\alpha_2\beta_2.$$

Finally in this section, we should mention the Inverse Function theorem, which will be needed in some subsequent chapters. This states that, if U is an open subset of $\mathbf{R}^n$ and $f : U \to \mathbf{R}^n$ is a smooth map, with the Jacobian matrix $J(f)$ non-singular at some point $P \in U$, then locally f is a homeomorphism from some open neighbourhood $V \ni P$ (with $V \subset U$) onto an open subset $V' \subset \mathbf{R}^n$, and that the inverse map $g : V' \to V$ is also smooth. We shall say that f is locally a *diffeomorphism*.

Most references for the Inverse Function theorem (for instance Theorem 9.24 of [11]) prove it in the form that, if f is continuously differentiable, then so too is the local inverse g. However, the Chain Rule then gives $J(g)$ as the inverse of $J(f)$, and so by Cramer's Rule, we may express the partial derivatives $\partial g/\partial y_j$ as rational functions of the partial derivatives $\partial f/\partial x_i$ of f. Thus, if f is also *smooth* (i.e. has partial derivatives of all orders), an inductive argument on the order shows that the same will be true for g.

4.2 Riemannian metrics on open subsets of **R**²

For notational simplicity, we now restrict ourselves to the case $n = 2$. Suppose that V is an open subset of $\mathbf{R}^2$, and let the standard coordinates on $\mathbf{R}^2$ be denoted by (u, v). A *Riemannian metric* on V is defined by giving *smooth* functions E, F, G on V, such that the matrix

$$\begin{pmatrix} E(P) & F(P) \\ F(P) & G(P) \end{pmatrix}$$

is positive definite for all $P \in V$. Thus (for $P \in V$), this determines an inner-product $\langle\,,\,\rangle_P$ on $\mathbf{R}^2$, where

$$\langle e_1, e_1 \rangle_P = E(P)$$

$$\langle e_1, e_2 \rangle_P = F(P)$$

$$\langle e_2, e_2 \rangle_P = G(P),$$

with e_1, e_2 denoting the standard basis for $\mathbf{R}^2$. A Riemannian metric should be thought of as a family of inner-products on the *tangent spaces* (all identified with $\mathbf{R}^2$).

The coordinate functions $u : V \to \mathbf{R}$, $v : V \to \mathbf{R}$ give rise to linear forms du and dv on $\mathbf{R}^2$, comprising the dual basis to the standard basis of $\mathbf{R}^2$, and hence also to *bilinear* forms du^2, $du\,dv$ and dv^2 on $\mathbf{R}^2$, where

$$du^2(\mathbf{h}, \mathbf{k}) = du(\mathbf{h})du(\mathbf{k}) \qquad\qquad \longleftrightarrow \quad \begin{pmatrix} 1 & 0 \\ 0 & 0 \end{pmatrix}$$

$$du\,dv(\mathbf{h}, \mathbf{k}) = \frac{1}{2}\,(du(\mathbf{h})dv(\mathbf{k}) + du(\mathbf{k})dv(\mathbf{h})) \qquad \longleftrightarrow \quad \begin{pmatrix} 0 & \frac{1}{2} \\ \frac{1}{2} & 0 \end{pmatrix}$$

$$dv^2(\mathbf{h}, \mathbf{k}) = dv(\mathbf{h})dv(\mathbf{k}) \qquad\qquad \longleftrightarrow \quad \begin{pmatrix} 0 & 0 \\ 0 & 1 \end{pmatrix}.$$

Here, the right-hand column indicates the 2×2 matrix defining the bilinear form.

Thus, the family of bilinear forms on $\mathbf{R}^2$ determined (for $P \in V$) by the matrix of smooth functions

$$\begin{pmatrix} E & F \\ F & G \end{pmatrix},$$

may be then written simply as $Edu^2 + 2Fdu\,dv + Gdv^2$. When the matrix is positive definite for all $P \in V$, this yields a smooth family of inner-products on $\mathbf{R}^2$, and this is precisely what we defined to be a *Riemannian metric* on V. Note that the Euclidean case (when the inner-product is the Euclidean inner-product for all $P \in V$) corresponds to the constant functions $E = G = 1$ and $F = 0$.

Given such a Riemannian metric on V, we may use this family of inner-products to define the length of *smooth* or *piecewise smooth* curves. For a smooth curve γ, we have its derivative, denoted γ' or $\dot{\gamma}$; we define $\|\gamma'\|$ at a point $\gamma(t)$ by means of the inner-product determined at $\gamma(t)$ by the given Riemannian metric. We shall refer to $\|\gamma'\|$ as the *speed* of γ at t.

Definition 4.1   For V an open subset of $\mathbf{R}^2$, equipped with a Riemannian metric $Edu^2 + 2Fdu\,dv + Gdv^2$, the *length* of a smooth curve

$$\gamma = (\gamma_1, \gamma_2) : [a, b] \to V$$

is defined to be

$$\int_a^b (E\dot{\gamma}_1^2 + 2F\dot{\gamma}_1\dot{\gamma}_2 + G\dot{\gamma}_2^2)^{1/2}\,dt.$$

Example   Consider the geometry of $V = \mathbf{R}^2$ with the Riemannian metric

$$\frac{4(du^2 + dv^2)}{(1 + u^2 + v^2)^2}.$$

This is an example of a metric which is *conformal* to the Euclidean metric, in that the inner-product at each point P is just a scaling of the Euclidean inner-product, the scaling being defined by a smooth function of P — in the above example, the function is $4/(1 + u^2 + v^2)^2$.

Let $\pi : S^2 \setminus \{N\} \to \mathbf{R}^2$ be the stereographic projection map. Given a point $P \in S^2 \setminus \{N\}$, we have $\pi(P) \in \mathbf{R}^2$, and the Riemannian metric defines an inner-product $\langle \, , \, \rangle_{\pi(P)}$ on $\mathbf{R}^2$. The tangent space to S^2 at P is defined to consist of vectors $\mathbf{x}$ such that $\mathbf{x} \cdot \overrightarrow{OP} = 0$. Note that, under this definition, the tangent space is considered as a real vector space with origin at P. This definition will be consistent with the general definition we give in Chapter 6 for the tangent space at a given point on any embedded surface in $\mathbf{R}^3$.

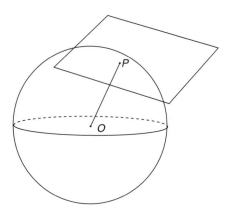

There is an inverse map $\sigma : \mathbf{R}^2 \to S^2 \setminus \{N\}$, given by

$$\sigma(u, v) = \left(2u/(1 + u^2 + v^2), 2v/(1 + u^2 + v^2), (u^2 + v^2 - 1)/(1 + u^2 + v^2)\right).$$

Considered as a map to $\mathbf{R}^3$, we know that σ is smooth. Let us consider the two partial derivatives of σ at $\pi(P)$, namely $\sigma_u(\pi(P)) = (d\sigma)_{\pi(P)}(e_1)$ and $\sigma_v(\pi(P)) = (d\sigma)_{\pi(P)}(e_2)$. On all of $\mathbf{R}^2$, we have that $\sigma(u, v) \cdot \sigma(u, v) = 1$; differentiating with respect to u and v, we deduce that $\sigma \cdot \sigma_u = 0$ and $\sigma \cdot \sigma_v = 0$ at all points of $\mathbf{R}^2$, and hence that both $\sigma_u(\pi(P))$ and $\sigma_v(\pi(P))$ are in the tangent space of S^2 at P. With σ as defined above, we check that

$$\sigma_u = \frac{2}{(u^2 + v^2 + 1)^2} \left(-u^2 + v^2 + 1, -2uv, -2u\right)$$

$$\sigma_v = \frac{2}{(u^2 + v^2 + 1)^2} \left(-2uv, u^2 - v^2 + 1, -2v\right).$$

We note that, when evaluated at any point of $\mathbf{R}^2$, the vectors σ_u and σ_v are non-zero and orthogonal, and therefore are linearly independent vectors in $\mathbf{R}^3$; thus for all $P \in S^2$, the derivative $(d\sigma)_{\pi(P)}$ induces an isomorphism of vector spaces between $\mathbf{R}^2$ and the tangent space to S^2 at P.

Given $\mathbf{x}_1, \mathbf{x}_2$ in the tangent space at P, we claim that

$$\mathbf{x}_1 \cdot \mathbf{x}_2 = \langle d\pi_P(\mathbf{x}_1), d\pi_P(\mathbf{x}_2) \rangle_{\pi(P)}.$$

Since $d\pi_P \circ (d\sigma)_{\pi(P)}$ is the identity on $\mathbf{R}^2$, this is equivalent to the statement that

$$(d\sigma)_{\pi(P)}\mathbf{u}_1 \cdot (d\sigma)_{\pi(P)}\mathbf{u}_1 = \langle \mathbf{u}_1, \mathbf{u}_2 \rangle_{\pi(P)}, \qquad (4.1)$$

for all $\mathbf{u}_1, \mathbf{u}_2 \in \mathbf{R}^2$. These equations say that the geometry on $\mathbf{R}^2$ induced from the Riemannian metric $4(du^2 + dv^2)/(1 + u^2 + v^2)^2$ corresponds to the standard spherical geometry on $S^2 \setminus \{N\}$. Thus for instance, one may check that the length of a semi-infinite ray starting at the origin is just π, since this is the length of the corresponding longitude on S^2.

To check that Equation (4.1) holds for all $\mathbf{u}_1, \mathbf{u}_2 \in \mathbf{R}^2$, we need only check (using the bilinearity) that it holds for the standard basis vectors e_1, e_2, which therefore reduces the problem to showing that

$$\sigma_u \cdot \sigma_u = \frac{4}{(u^2 + v^2 + 1)^2}, \quad \sigma_u \cdot \sigma_v = 0 \quad \text{and} \quad \sigma_v \cdot \sigma_v = \frac{4}{(u^2 + v^2 + 1)^2},$$

at all points of $\mathbf{R}^2$. To check these is an elementary manipulation.

The basic theory of this example will be generalized in Chapter 6 to arbitrary embedded surfaces in $\mathbf{R}^3$.

4.3   Lengths of curves

In this section, V will denote an open subset of $\mathbf{R}^2$, equipped with a Riemannian metric $E\,du^2 + 2F\,du\,dv + G\,dv^2$. Given a smooth curve $\gamma = (\gamma_1, \gamma_2) : [a, b] \to V$, its length was defined as

$$\int_a^b \|\gamma'\| \, dt = \int_a^b (E\dot{\gamma}_1^2 + 2F\dot{\gamma}_1\dot{\gamma}_2 + G\dot{\gamma}_2^2)^{1/2} \, dt.$$

Lemma 4.2   *Given a curve γ as above, its length is invariant under reparametrizations of the curve given by smooth functions $f : [\tilde{a}, \tilde{b}] \to [a, b]$ with $f'(s) > 0$ for all $s \in [\tilde{a}, \tilde{b}]$.*

Proof   We let $\tilde{\gamma} : [\tilde{a}, \tilde{b}] \to V$ be defined by $\tilde{\gamma}(s) = \gamma(f(s))$ for $s \in [\tilde{a}, \tilde{b}]$. We need to prove that length $\tilde{\gamma}$ = length γ. The Chain Rule implies that $\tilde{\gamma}'(s) = f'(s)\,\gamma'(f(s))$, and hence, taking norms at $\gamma(f(s)) = \tilde{\gamma}(s)$ and using $f'(s) > 0$, that $\|\tilde{\gamma}'(s)\| = |f'(s)|\,\|\gamma'(f(s))\| = f'(s)\,\|\gamma'(f(s))\|$. Using the change of variable formula for integrals, with the change of variable $t = f(s)$, we deduce that

$$\text{length } \tilde{\gamma} = \int_{a'}^{b'} \|\tilde{\gamma}'(s)\| \, ds$$

$$= \int_{a'}^{b'} f'(s) \, \|\gamma'(f(s))\| \, ds$$

$$= \int_a^b \|\gamma'(t)\| \, dt = \text{length } \gamma. \qquad \square$$

Suppose $\gamma : [a, b] \to V$ is a unit speed smooth curve, i.e. $\|\gamma'(t)\| = 1$ for all t; then on integrating, we obtain $t = a + s(t)$, where $s(t)$ is the length of the curve $\gamma|_{[a,t]}$. Conversely, if this relation holds for all t, then on differentiating, we see that γ is a unit speed curve. A curve therefore has unit speed if and only if it is parametrized, up to an additive constant, by the arc-length.

Lemma 4.3 *Given a smooth curve $\gamma : [a, b] \to V$ of length l and with nowhere vanishing derivative, we can find a smooth reparametrization given by $f : [0, l] \to [a, b]$ such that $\tilde{\gamma} = \gamma \circ f$ has unit speed.*

Proof   We reparametrize γ in terms of arc-length. Let

$$g(t) = \int_a^t \|\gamma'(t)\| \, dt$$

be the arc-length function. This is a strictly increasing function $g : [a, b] \to [0, l]$; moreover, as $\|\gamma'(t)\|^2$ is clearly a smooth function of t, our assumptions imply that the same is true of both $\|\gamma'(t)\|$ and $g(t)$. Let $f : [0, l] \to [a, b]$ be the inverse function to g. Since $f \circ g$ is the identity, we deduce, for $s = g(t)$, that

$$\frac{df}{ds}(s) \frac{dg}{dt}(t) = 1;$$

in particular, we see that f is a smooth function of s. Differentiating the above integral however gives $dg/dt = \|\gamma'(t)\|$. Thus $(df/ds)(s) = 1/\|\gamma'(t)\| > 0$ for all s.

Applying the Chain Rule, as in Lemma 4.2, we have

$$\|\tilde{\gamma}'(s)\| = |f'(s)| \, \|\gamma'(t)\| = 1$$

as required. $\square$

Remark   Smooth curves with nowhere vanishing derivative are called *smoothly immersed* curves. An example of a smooth curve which is not smoothly immersed is given by the cuspidal cubic $\gamma : [-1, 1] \to \mathbf{R}^2$ given by $\gamma(t) = (t^2, t^3)$; here $\gamma'(0) = 0$. In fact, one may easily check in this case that no (continuous) reparametrization of γ is smoothly immersed (Exercise 4.2).

We now show that any Riemannian metric on a connected open subset of $\mathbf{R}^2$ gives rise to an associated metric (which is in fact intrinsic). As is the case for the other results of this section, this will be equally true for a general abstract surface equipped with a Riemannian metric, once we have defined the terms in Chapter 8, and the proofs go over without change to the more general case.

For the present however, we let V be an open connected subset of $\mathbf{R}^2$, equipped with a Riemannian metric. This enables us to define the lengths of curves in V. We can define a distance function ρ (called the *Riemannian distance*) by defining $\rho(P, Q)$ to be the infimum of the lengths of piecewise smooth curves joining P to Q – by Exercise 1.7, such curves exist. This is a metric on V; it is clear that this distance is non-negative and symmetric, and the triangle inequality is an easy exercise. The only axiom for a metric that is not entirely straightforward is covered by the following lemma.

Lemma 4.4   *With the distance function ρ defined on V as above, we have*
$\rho(P, Q) > 0$ *for* $P \neq Q$ *in* V.

Proof   The fact that $\rho(P, Q) \geq 0$ is obvious, but the strict inequality needs proof. The idea is to compare our Riemannian metric locally with an appropriate multiple of the Euclidean metric. We let d denote the Euclidean metric on V.

We recall that a 2×2 real symmetric matrix (a_{ij}) is positive definite if and only if $a_{11} > 0$ and $a_{11}a_{22} > a_{12}^2$. Since the matrix

$$\begin{pmatrix} E(P) & F(P) \\ F(P) & G(P) \end{pmatrix}$$

is positive definite, the same will therefore be true for the matrix

$$\begin{pmatrix} E(P) - \varepsilon^2 & F(P) \\ F(P) & G(P) - \varepsilon^2 \end{pmatrix}$$

for some sufficiently small $\varepsilon > 0$. Moreover, the matrix

$$\begin{pmatrix} E(P') - \varepsilon^2 & F(P') \\ F(P') & G(P') - \varepsilon^2 \end{pmatrix}$$

will remain positive definite for all P' in some Euclidean ball $B(P, \delta) \subset V$, of Euclidean radius δ. Thus for any $P' \in B(P, \delta)$ and any $\mathbf{x} = (x_1, x_2)^t \in \mathbf{R}^2$, we have

$$\langle \mathbf{x}, \mathbf{x} \rangle_{P'} := E(P')x_1^2 + 2F(P')x_1x_2 + G(P')x_2^2 \geq \varepsilon^2(x_1^2 + x_2^2).$$

For any piecewise smooth curve γ in $B(P, \delta)$, it follows therefore from the definition of lengths that the length of γ with respect to the Riemannian metric is at least ε times the length of γ in the Euclidean metric.

Given now $P \neq Q \in V$, consider any piecewise smooth curve $\gamma : [a, b] \to V$ joining the two points. Suppose that γ is not contained in the ball $B(P, \delta)$. There exists $\xi \in [a, b]$ such that $\gamma|_{[a, \xi)}$ is contained in $B(P, \delta)$ but $\gamma(\xi)$ lies on the boundary of the ball. The above argument then gives that

$$\text{length } \gamma \geq \text{length } \gamma|_{[a, \xi)} \geq \varepsilon \delta,$$

since the Euclidean length of $\gamma|_{[a, \xi)}$ is at least δ. If however γ is contained in the ball, then the above argument gives length $\gamma \geq \varepsilon d(P, Q)$. Therefore, taking the infimum over all such γ,

$$\rho(P, Q) \geq \varepsilon \min\{\delta, d(P, Q)\} > 0.$$

$\square$

Example   Let us consider $\mathbf{R}^2$ equipped with a Riemannian metric

$$dx^2/(1+x^2)^2 + dy^2/(1+y^2)^2.$$

We show that, in the associated metric defined as above, the distances are bounded above by 2π (*not* claimed to be the *optimal* bound).

To see this, we show that both horizontal and vertical line segments (parametrized linearly) have lengths bounded above by π. Since any two points of $\mathbf{R}^2$ may be joined by a curve consisting at most one horizontal line segment and at most one vertical line segment, therefore having total length less than 2π, the result follows.

Suppose therefore $\gamma : [0, 1] \to \mathbf{R}^2$ is a horizontal line segment, given by $\gamma(t) = (at + b, 0)$. We may assume (by considering, if necessary, the opposite curve $-\gamma$ instead) that $a > 0$. Hence

$$\text{length } \gamma = \int_0^1 \frac{a}{1 + (at+b)^2} \, dt = \int_b^{a+b} \frac{ds}{1 + s^2} = \left[\tan^{-1} s \right]_a^{a+b},$$

using the substitutions $s = at + b$ and $u = \tan^{-1} s$. Since $|\tan^{-1} s| < \pi/2$ for all $s \in \mathbf{R}$, we deduce that length $\gamma < \pi$. Similarly, any vertical line segment has length bounded above by π.

4.4   Isometries and areas

Suppose $\phi : \tilde{V} \to V$ is a diffeomorphism (that is, a smooth map with a smooth inverse) between open subsets of $\mathbf{R}^2$. Suppose we have Riemmanian metrics on $\tilde{V}$ and V giving rise to families of inner-products on $\mathbf{R}^2$, say $\langle \, , \, \rangle_P^{\tilde{}}$ for $P \in \tilde{V}$ and $\langle \, , \, \rangle_Q$ for $Q \in V$.

Definition 4.5   A diffeomorphism ϕ is called an *isometry* if for all $P \in \tilde{V}$

$$\langle \mathbf{x}, \mathbf{y} \rangle_P^{\tilde{}} = \langle d\phi_P(\mathbf{x}), d\phi_P(\mathbf{y}) \rangle_{\phi(P)} \quad \text{for all } \mathbf{x}, \mathbf{y} \in \mathbf{R}^2.$$

Equivalently, in terms of coordinates, given any $\mathbf{x}, \mathbf{y} \in \mathbf{R}^2$, we have

$$\mathbf{x}^t \begin{pmatrix} \tilde{E} & \tilde{F} \\ \tilde{F} & \tilde{G} \end{pmatrix}_P \mathbf{y} = \mathbf{x}^t J^t \begin{pmatrix} E & F \\ F & G \end{pmatrix}_{\phi(P)} J \mathbf{y},$$

with $J = J(\phi)$ the Jacobian matrix. This in turn is equivalent to the condition that

$$\begin{pmatrix} \tilde{E} & \tilde{F} \\ \tilde{F} & \tilde{G} \end{pmatrix} = J^t \begin{pmatrix} E \circ \phi & F \circ \phi \\ F \circ \phi & G \circ \phi \end{pmatrix} J,$$

as matrices of functions on $\tilde{V}$.

If $\tilde{\gamma} : [0, 1] \to \tilde{V}$ is a smooth curve, then $\gamma = \phi \circ \tilde{\gamma} : [0, 1] \to V$ is also smooth. Using the Chain Rule and letting $P = \tilde{\gamma}(t)$, we have

$$\langle \gamma'(t), \gamma'(t) \rangle_{\gamma(t)} = \langle d\phi_P(\tilde{\gamma}'(t)), d\phi_P(\tilde{\gamma}'(t)) \rangle_{\phi(P)}$$
$$= \langle \tilde{\gamma}'(t), \tilde{\gamma}'(t) \rangle_{\tilde{\gamma}(t)}$$

when ϕ is an isometry. Therefore, if ϕ is an isometry, then

$$\text{length } \tilde{\gamma} = \text{length } \gamma$$
$$= \int_0^1 \langle \gamma'(t), \gamma'(t) \rangle_{\gamma(t)}^{1/2} \, dt.$$

So ϕ preserves the lengths of curves (and therefore also distances in the associated metric, defined above). Thus, if ϕ is an isometry in the above sense, it is also an isometry of the corresponding metric spaces.

We now introduce the concept of *area*.

Definition 4.6   Given a Riemannian metric $E \, du^2 + 2F \, du \, dv + G \, dv^2$ on an open subset $V \subset \mathbf{R}^2$, and a region $W \subset V$, we define its *area* (with respect to the metric) to be

$$\int_W (EG - F^2)^{1/2} \, du \, dv,$$

when the integral is defined, using the standard notation for integrals over regions in the plane.

Example   If we consider $\mathbf{R}^2$ equipped with the metric from the example of the previous section, then $\mathbf{R}^2$ has area π^2 with respect to this metric. Applying the above definition, the area of $\mathbf{R}^2$ is given by the integral

$$\int_{-\infty}^{\infty} \int_{-\infty}^{\infty} \frac{dx \, dy}{(1 + x^2)(1 + y^2)} = \left(\int_{-\infty}^{\infty} \frac{du}{1 + u^2} \right)^2 = \pi^2.$$

The reader will not be surprised to learn that isometries also preserve areas.

Proposition 4.7   *Suppose that V and $\tilde{V}$ are open subsets of $\mathbf{R}^2$ equipped with Riemannian metrics, and that $\phi : \tilde{V} \to V$ is an isometry. For any region $W \subset V$ for which the area exists, the region $\phi^{-1}W$ of $\tilde{V}$ has the same area as W.*

Proof   Since ϕ is an isometry, we have

$$\begin{pmatrix} \tilde{E} & \tilde{F} \\ \tilde{F} & \tilde{G} \end{pmatrix}_P = J^t \begin{pmatrix} E & F \\ F & G \end{pmatrix}_{\phi(P)} J$$

for $P \in \tilde{V}$, where $J = J(\phi)$ is the Jacobian matrix representing $d\phi_P$.

The change of variables formula for integrals on $\mathbf{R}^2$ ([11], Theorem 10.9), implies that for any continuous function H on $W \subset V$, we have

$$\int_W H \, du \, dv = \int_{\phi^{-1}W} (H \circ \phi) \, |\det J(\phi)| \, d\tilde{u} \, d\tilde{v}.$$

Setting

$$H = (EG - F^2)^{1/2} \text{ on } V, \quad \text{and} \quad \tilde{H} = (\tilde{E}\tilde{G} - \tilde{F}^2)^{1/2} \text{ on } \tilde{V},$$

we have (taking determinants in the above formula) that

$$\tilde{H} = (H \circ \phi) \, |\det J(\phi)|.$$

Therefore,

$$\text{area}(W) = \int_W H \, du \, dv = \int_{\phi^{-1}W} \tilde{H} \, d\tilde{u} \, d\tilde{v} = \text{area}(\phi^{-1}W),$$

as claimed. $\qquad\qquad\qquad\qquad\qquad\qquad\qquad\qquad\qquad\qquad\qquad\quad$ $\square$

Exercises

4.1 Suppose that U is an open subset of $\mathbf{R}^2$ and $f : U \to \mathbf{R}^2$ is a smooth map with the property that, for all points $P \in U$, the linear maps df_P are non-singular and preserve angles (and orientations) between non-zero vectors — such a map is called *conformal*. Show that f is a complex analytic function, when considered as a map from U to $\mathbf{C}$.

4.2 Let $\gamma : [-1, 1] \to \mathbf{R}^2$ be the smooth plane curve given by $\gamma(t) = (t^3, t^6)$. Find a homeomorphism $f : [a, b] \to [-1, 1]$ (for $[a, b]$ a suitable real interval) with the property that the (continuous) curve $\eta = \gamma \circ f$ is a smoothly immersed curve.

Now let γ be the smooth plane curve given by $\gamma(t) = (t^2, t^3)$; prove that, in this case, there does not exist such a homeomorphism f with $\eta = \gamma \circ f$ a smoothly immersed curve.

4.3 We define a Riemannian metric on the unit disc $D \subset \mathbf{R}^2$ by

$$(du^2 + dv^2)/(1 - (u^2 + v^2)).$$

Show that, in the corresponding metric on D (defined by taking the infima of lengths of curves between points), the distances are bounded, but that the areas are unbounded.

4.4 Consider the punctured disc $D^* = D \setminus \{0\} \subset \mathbf{R}^2$ equipped with a Riemannian metric $(du^2 + dv^2)/(u^2 + v^2)^a$, for some real number a. For which values of a are the distances in this metric bounded? For which values of a does D^* have finite area?

4.5 Given an open subset $U \subset \mathbf{R}^2$, a surface S in $U \times \mathbf{R}$ is defined by the equation $z = h(x, y)$, for some smooth real-valued function h on U. We represent the points (x, y, z) of S by their projections (x, y) in U, and we wish to equip U with a Riemannian metric $E \, dx^2 + 2F \, dx \, dy + G \, dy^2$, so that the length of any curve on S is just the length (with respect to the given metric) of the corresponding curve on U. Find formulae for the required smooth functions E, F and G on U. If the area of some connected open set $W \subset U$ with respect this metric is the same as the Euclidean area of W, show that the function h is constant on W.

4.6   We let $V \subset \mathbf{R}^2$ denote the square given by $|u| < 1$ and $|v| < 1$, and define two Riemannian metrics on V given by

$$du^2/(1-u^2)^2 + dv^2/(1-v^2)^2, \quad \text{and} \quad du^2/(1-v^2)^2 + dv^2/(1-u^2)^2.$$

Prove that there is no isometry between the two spaces, but that an area-preserving diffeomorphism does exist.

4.7   Show that $\mathbf{R}^2$ equipped with the Riemannian metric

$$(1+u^2)du^2 + 2u \, du \, dv + dv^2$$

is isometric to $\mathbf{R}^2$ with the Euclidean metric.

4.8   Consider $\mathbf{R}^2 \setminus \{0\}$ equipped with the Riemannian metric

$$du^2/u^2 + dv^2/v^2.$$

Show that the map given by $(u, v) \mapsto (u, 1/v)$ is an isometry with respect to this metric. Exhibit a non-abelian group of order 32 consisting of isometries of the metric.

4.9   Consider the Riemannian metric on $\mathbf{R}^2$ given by

$$4(du^2 + dv^2)/(1 + u^2 + v^2)^2.$$

By calculating explicitly on $\mathbf{R}^2$, show that a circle centred at the origin with Euclidean radius $\cot(\rho/2)$ has circumference $2\pi \sin \rho$ and area $2\pi(1 - \cos \rho)$.

4.10   Consider $\mathbf{R}^2 \setminus \{0\}$ equipped with the Riemannian metric from the previous exercise. Let $\phi : \mathbf{R}^2 \setminus \{0\} \to \mathbf{R}^2 \setminus \{0\}$ be given by

$$\phi(u, v) = \left(\frac{-u}{u^2 + v^2}, \frac{v}{u^2 + v^2} \right).$$

Verify directly that ϕ is an isometry, and then give a geometric explanation of this fact.

5 Hyperbolic geometry

In the previous chapter, we introduced the notion of a Riemannian metric on an open subset of $\mathbf{R}^2$. In this chapter, we study two particular examples of this, namely the Poincaré disc and upper half-plane models of the hyperbolic plane. These models are isometric to each other, and also to the hyperboloid model of the hyperbolic plane, which we study at the end of the chapter. The hyperbolic plane represents the third standard type of geometry, after Euclidean and spherical, and as such occupies a central role in geometry. Studying the hyperbolic plane via the Poincaré models facilitates many explicit calculations; in the context of this book it will also provide a useful illustration for the general theory of Riemannian metrics.

5.1 Poincaré models for the hyperbolic plane

We recall from the last chapter that the spherical metric on $S^2 \setminus \{N\}$ could be interpreted, via stereographic projection, in terms of the Riemannian metric on $\mathbf{R}^2$ given by $4(du^2 + dv^2)/(1 + u^2 + v^2)^2$. The disc model of the hyperbolic plane can be defined by analogy to this by changing a sign.

Definition 5.1   The *disc model* for a hyperbolic space is defined on the unit disc $D \subset \mathbf{C} = \mathbf{R}^2$, where $D = \{\zeta : |\zeta| < 1\}$, with a Riemannian metric given by

$$\frac{4(du^2 + dv^2)}{(1 - u^2 - v^2)^2},$$

where $\zeta = u + iv$. Often the metric is written as

$$\frac{4\,|d\zeta|^2}{(1 - |\zeta|^2)^2},$$

where *either* one takes $|d\zeta|^2 = du^2 + dv^2$ as a formal definition *or* one interprets $d\zeta = du + idv$ and $d\bar{\zeta} = du - idv$ as real linear forms $\mathbf{C} \to \mathbf{C}$, with $|d\zeta|^2$ being the real bilinear form $d\zeta\, d\bar{\zeta} = du^2 + dv^2$.

The Riemannian metric is just a scaling of the Euclidean metric by a factor $4/(1 - r^2)^2$. In the notation of the previous chapter, we have

$$E = G = 4/(1 - r^2)^2 \quad \text{and} \quad F = 0.$$

Geometrically, we have that distances at radius r are locally scaled by $2/(1 - r^2)$, and areas by $(EG - F^2)^{1/2} = 4/(1 - r^2)^2$.

The *upper half-plane* $H = \{z \in \mathbf{C} : \operatorname{Im} z > 0\}$ is conformally equivalent to D

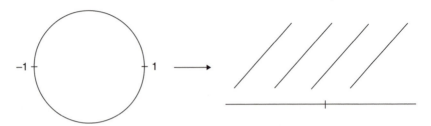

via the Möbius transformation

$$\zeta \mapsto \frac{i(1 + \zeta)}{1 - \zeta}.$$

It is an easy check (Exercise 5.1) that this may be defined by an element of $SU(2)$, and hence it corresponds, under stereographic projection, to a rotation of S^2, sending the lower hemisphere to the vertical hemisphere at the back.

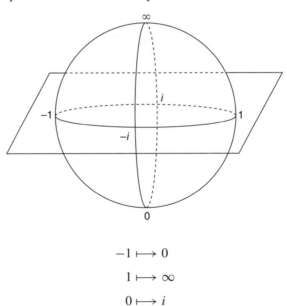

$$-1 \longmapsto 0$$
$$1 \longmapsto \infty$$
$$0 \longmapsto i$$

We use z now for the complex coordinate on H, with $z = x + iy$; the above transformation is then $z = \frac{i(1+\zeta)}{1-\zeta}$, with inverse $\zeta = \frac{z-i}{z+i}$. Via the derivative of this map, the Riemannian metric on D induces a Riemannian metric on H, which we now calculate.

The Euclidean inner-product on $\mathbf{R}^2 = \mathbf{C}$ may be written as

$$\langle w_1, w_2 \rangle = \mathrm{Re}(w_1 \bar{w}_2) = \frac{1}{2}(w_1 \bar{w}_2 + \bar{w}_1 w_2).$$

Therefore, at a given point $z \in H$, the inner-product on $\mathbf{R}^2 = \mathbf{C}$ induced from the Euclidean inner-product at $\zeta = \frac{z-i}{z+i}$ is

$$\langle w_1, w_2 \rangle_z = \left\langle \frac{d\zeta}{dz} w_1, \frac{d\zeta}{dz} w_2 \right\rangle_{\mathrm{Euclidean}}$$

$$= \left| \frac{d\zeta}{dz} \right|^2 \mathrm{Re}(w_1 \bar{w}_2)$$

i.e. we get the Riemannian metric $\left| \frac{d\zeta}{dz} \right|^2 (dx^2 + dy^2)$ on H. Thus the Riemannian metric $|d\zeta|^2$ on D corresponds to the Riemannian metric $\left| \frac{d\zeta}{dz} \right|^2 |dz|^2$ on H. We now calculate

$$\frac{d\zeta}{dz} = \frac{1}{z+i} - \frac{z-i}{(z+i)^2} = \frac{2i}{(z+i)^2}.$$

Also

$$1 - |\zeta|^2 = 1 - \frac{|z-i|^2}{|z+i|^2},$$

and so

$$\frac{1}{1 - |\zeta|^2} = \frac{|z+i|^2}{|z+i|^2 - |z-i|^2} = \frac{|z+i|^2}{4 \,\mathrm{Im}\, z}.$$

Therefore the metric on H corresponding to $4 |d\zeta|^2 / (1 - |\zeta|^2)^2$ on D is just

$$4 \frac{4}{|z+i|^4} \left(\frac{|z+i|^2}{4 \,\mathrm{Im}\, z} \right)^2 |dz|^2 = \frac{|dz|^2}{(\mathrm{Im}\, z)^2}$$

$$= \frac{dx^2 + dy^2}{y^2}.$$

This is again just a scaling of the Euclidean metric — lengths are locally scaled by $1/y$ and areas by $1/y^2$.

We have constructed this metric on H precisely so that the given conformal map between D and H is an *isometry* of Riemannian metrics. The spaces D and H are called, respectively, the Poincaré disc and upper half-plane models for the hyperbolic plane. The fact that they are isometric means that, from the point of view of the geometry, we can switch between the two models at will, choosing whichever is most convenient for the particular problem we wish to study.

5.2   Geometry of the upper half-plane model H

Consider the group $PSL(2, \mathbf{R})$ of Möbius transformations

$$z \mapsto \frac{az + b}{cz + d}$$

with $a, b, c, d \in \mathbf{R}$ and $\det\left(\begin{smallmatrix} a & b \\ c & d \end{smallmatrix}\right) = 1$. It is easy to check that these are precisely the Möbius transformations of $\mathbf{C}_\infty$ which send $\mathbf{R} \cup \{\infty\}$ to $\mathbf{R} \cup \{\infty\}$ and send H to H. A Möbius transformation with *real* coefficients may always be represented by a *real* matrix with determinant ± 1; the condition that the determinant is positive is just saying that the upper half-plane is sent to itself (and not to the lower half-plane).

Proposition 5.2   *The elements of $PSL(2, \mathbf{R})$ are isometries on H, and hence preserve the lengths of curves.*

Proof   Recall that $PSL(2, \mathbf{R})$ is generated by the elements

$$z \mapsto z + a \qquad (a \in \mathbf{R}) \qquad \text{(translations)}$$

$$z \mapsto az \qquad (a \in \mathbf{R}_+) \qquad \text{(dilations)}$$

$$z \mapsto -\frac{1}{z}$$

Claim   Each of these preserves the metric $|dz|^2 / y^2$.

The first two are clear — we check the third. Set $w = -1/z$.
Since $dw/dz = 1/z^2$, the induced map on $\mathbf{C} = \mathbf{R}^2$ is given as multiplication by $1/z^2$. Under this linear map, the Euclidean metric $|dw|^2$ at w corresponds to $|dz|^2 / |z|^4$ at z. Note however that

$$\text{Im}(w) = \text{Im}(-1/z) = -\frac{1}{|z|^2}\, \text{Im}\,\bar{z} = \frac{\text{Im}\, z}{|z|^2}.$$

Thus

$$\frac{|dw|^2}{\text{Im}(w)^2} = \frac{|dz|^2 / |z|^4}{\text{Im}(z)^2 / |z|^4} = \frac{|dz|^2}{(\text{Im}\, z)^2},$$

as required, and the map is an isometry.   □

We shall see later (Remark 5.17) that $PSL(2, \mathbf{R})$ is in fact an index two subgroup of the full isometry group.

Remark 5.3   Since $PSL(2, \mathbf{R})$ contains the Möbius transformations of the form $z \mapsto az + b$, with $a > 0$, it acts transitively on H, i.e. for any points $z_1, z_2 \in H$, there exists $g \in PSL(2, \mathbf{R})$ with $g(z_1) = z_2$.

Recall from Chapter 2 that any Möbius transformation on $\mathbf{C}$ sends circles and straight lines to circles and straight lines. Since it is analytic, it also preserves angles, by the argument in Section 4.1. So if L is the imaginary axis, and $g \in PSL(2, \mathbf{R})$, we deduce that $g(L)$ is a circle or a straight line, orthogonal to the real axis, since g sends $\mathbf{R} \cup \{\infty\}$ to itself.

Therefore, if $L^+ := \{it : t > 0\}$, then $g(L^+)$ is either a vertical half-line, or a semicircle (whose ends are on the real axis). We call these the *hyperbolic lines* in H. We note that two distinct hyperbolic lines meet in at most one point.

Lemma 5.4   *Through any two points $z_1, z_2 \in H$, there exists a unique hyperbolic line l.*

Proof   This is clearly true if $\operatorname{Re} z_1 = \operatorname{Re} z_2$. Suppose then $\operatorname{Re} z_1 \neq \operatorname{Re} z_2$. We can locate the centre of the required semicircle by the construction as shown of a perpendicular bisector, and hence l is uniquely determined.

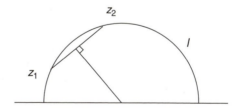

$\square$

Lemma 5.5   *$PSL(2, \mathbf{R})$ acts transitively on the set of hyperbolic lines.*

Proof   We show that for any hyperbolic line l, there exists $g \in PSL(2, \mathbf{R})$ such that $g(l) = L^+$, from which the general statement follows.

This is clear if l is a vertical line. If l is a semicircle with end-points $s < t \in \mathbf{R}$, then we take

$$g(z) = \frac{z - t}{z - s}$$

(noting that $\det\left(\begin{smallmatrix} 1 & -t \\ 1 & -s \end{smallmatrix}\right) = t - s > 0$).

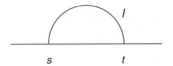

We observe that $g(t) = 0$, $g(s) = \infty$ and so $g(l) = L^+$.   □

Remark 5.6   If, in the proof of Lemma 5.5, we compose g with $z \mapsto -\frac{1}{z}$, we obtain a Möbius transformation h with $h(s) = 0$, $h(t) = \infty$. We can also ensure (scaling by a real number) that a given point $P \in l$ goes to i, say. Thus $PSL(2, \mathbf{R})$ in fact acts transitively on pairs (l, P) consisting of a hyperbolic line l and a point $P \in l$.

Definition 5.7   We now consider the metric defined by the Riemannian metric on H, where the distance between any two points is the infimum of the lengths of piecewise smooth curves joining the points; we shall call this the *hyperbolic distance ρ*. So Proposition 5.2 implies that $PSL(2, \mathbf{R})$ preserves hyperbolic distance.

Given points $z_1, z_2 \in H$, there is a unique hyperbolic line through z_1 and z_2; we let z_1^*, z_2^* be as shown in the diagram (possibly with $z_2^* = \infty$, if the hyperbolic line is a vertical half-line).

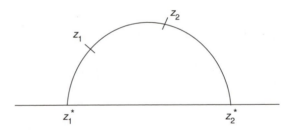

As we argued above, there exists an element $h \in PSL(2, \mathbf{R})$ with $h(z_1^*) = 0$ and $h(z_2^*) = \infty$, which therefore sends the hyperbolic line through z_1 and z_2 to the positive imaginary axis L^+. Therefore, $h(z_1) = iu$, $h(z_2) = iv$ with $u < v$. Since h preserves distances, $\rho(z_1, z_2) = \rho(iu, iv)$.

Let us therefore consider the case when the points are of the form $z_1 = iu$, $z_2 = iv$, with $u < v$. Suppose $\tau : [0, 1] \rightarrow H$ is a piecewise smooth curve such that $\tau(t) = i f(t) \in L^+$ for all t, with $\tau(0) = iu$, $\tau(1) = iv$.

We say that a hyperbolic line segment $\gamma : [0, 1] \rightarrow H$ is *monotonically parametrized* if $\rho(\gamma(0), \gamma(t))$ is a monotonic function of t. This property is clearly preserved when we take the image $h \circ \gamma$ of γ under an element h of $PSL(2, \mathbf{R})$. In the specific case under consideration, τ being monotonically parametrized is equivalent to the (piecewise smooth) function f being monotonic increasing. By the Mean Value theorem, this is equivalent to the condition that $f'(t) \geq 0$, whenever the derivative exists and is continuous.

Suppose now that τ is monotonically parametrized; then

$$\text{length } \tau = \int_0^1 \frac{|df/dt|}{f} \, dt$$

$$= \int_0^1 \frac{df/dt}{f} \, dt$$

$$= \log \frac{v}{u}.$$

We claim that this is $\rho(z_1, z_2)$.

Proposition 5.8   *Suppose z_1, z_2 are points of H, and $\gamma : [0,1] \to H$ is a piecewise smooth curve from z_1 to z_2. Then* length $\gamma \geq \rho(z_1, z_2)$, *with equality if and only if γ is a monotonic parametrization of the hyperbolic line segment $[z_1, z_2]$.*

Proof   As argued above, we may reduce to the case when $z_1 = iu$, $z_2 = iv$, with $u < v$. Suppose $\gamma = \gamma_1 + i\gamma_2 : [0,1] \to H$ is a piecewise smooth curve with $\gamma(0) = iu$, $\gamma(1) = iv$. Then

$$\text{length } \gamma = \int_0^1 \left(\left(\frac{d\gamma_1}{dt}\right)^2 + \left(\frac{d\gamma_2}{dt}\right)^2 \right)^{1/2} \frac{dt}{\gamma_2(t)}$$

$$\geq \int_0^1 \left| \frac{d\gamma_2}{dt} \right| \frac{dt}{\gamma_2(t)}$$

$$\geq \int_0^1 \frac{d\gamma_2/dt}{\gamma_2} \, dt$$

$$= [\log \gamma_2]_0^1 = \log \frac{v}{u},$$

with equality if and only if $\frac{d\gamma_1}{dt} = 0$ and $\frac{d\gamma_2}{dt} \geq 0$ wherever they exist and are continuous, i.e. if and only if $\gamma_1 \equiv 0$ and γ_2 is monotonic.   □

Thus the hyperbolic distance between two points $z_1, z_2 \in H$ is just the length of the unique hyperbolic line segment $[z_1, z_2]$ between them. Moreover, if a hyperbolic line segment $\gamma : [0,1] \to H$ is monotonically parametrized, then $\rho(\gamma(0), \gamma(t)) =$ length $\gamma|_{[0,t]}$ for all $t \in [0,1]$.

Remark 5.9   For a general continuous curve $\gamma : [0,1] \to H$ with $\gamma(0) = z_1$, $\gamma(1) = z_2$, we can define its length (when it exists) by taking dissections $\mathcal{D} : 0 = t_0 < t_1 < \cdots < t_N = 1$; we set $P_i = \gamma(t_i)$ and $\tilde{s}_\mathcal{D} = \sum_{i=1}^N \rho(P_{i-1}, P_i)$, and we define length $\gamma = \sup_\mathcal{D} \tilde{s}_\mathcal{D}$. It is now a formal consequence of the triangle inequality (exactly as in the proof of Proposition 2.10) that length $\gamma \geq \rho(z_1, z_2)$, with equality if and only if γ is a monotonic parametrization of the hyperbolic line segment $[z_1, z_2]$.

5.3   Geometry of the disc model D

We have *isometries* between the two Poincaré models:

$$
\begin{array}{cc}
D \to H & H \to D \\
z = \frac{i(1+\zeta)}{1-\zeta} & \zeta = \frac{z-i}{z+i}
\end{array}
$$

Recalling that under any Möbius transformation, circles and straight lines are mapped to circles and straight lines and angles are preserved, we deduce the following facts.

(i)   The Möbius transformations sending the unit circle to itself and D to D correspond on H to Möbius transformations sending the real line to itself and H to H, i.e. the elements of $PSL(2, \mathbf{R})$ acting on H, which we saw were isometries. Since the correspondence between D and H is itself an isometry (by construction), the Möbius transformations of D to itself are isometries of D — they form a group G.

(ii)   Hyperbolic lines in D are given by circle segments orthogonal to the unit circle, including diameters.

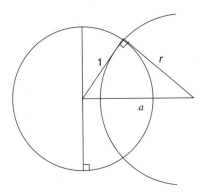

If the circular arc has centre at $a > 0 \in \mathbf{R}$ and radius $r > 0$, then by Pythogoras $a^2 = r^2 + 1$; we therefore deduce the geometric facts that $a > 1$ (the centre of the circular arc lies outside the unit circle) and $r < a$ (that the corresponding circle does not contain 0). In particular, a hyperbolic line in D contains the origin if and only if it is a diameter of the unit disc.

(iii) G acts transitively on the set of pairs (l, P), consisting of a hyperbolic line $l \subset D$ and a point $P \in l$ — cf. Lemma 5.5. Given a point P on a hyperbolic line l, we may apply an element of G which sends P to the origin (and therefore l to a diameter). Given an angle α, it is then clear that there exists a unique hyperbolic line l' through P at an angle α (in a given orientation) to l — when $P = 0$, this will be another diameter. We recall here that elements of G preserve angles and orientations.

(iv)   The curves of minimum length between two given points on D correspond to hyperbolic line segments.

We also denote the hyperbolic distance on D by ρ, and we now denote by z the coordinate on D.

Lemma 5.10 *(i) Rotations $z \mapsto e^{i\theta}z$ are elements of G. (ii) If $a \in D$, then $z \mapsto g(z) = \frac{z-a}{1-\bar{a}z}$ is an element of G.*

Proof (i) Clear. (ii) Observe that g sends the unit circle to itself (since if $|z| = 1$, then

$$|1 - \bar{a}z| = |\bar{z}(1 - \bar{a}z)| = |\bar{z} - \bar{a}| = |z - a|,$$

i.e. $|g(z)| = 1$). Since $g(a) = 0$, the claim follows. $\qquad \square$

Remark 5.11 In fact, by Exercise 5.6, any element of G is of the form

$$z \mapsto e^{i\theta}\left(\frac{z-a}{1-\bar{a}z}\right).$$

Proposition 5.12 *If $0 \le r < 1$, then*

$$\rho(0, re^{i\theta}) = \rho(0, r) = 2\tanh^{-1} r.$$

In general, given $z_1, z_2 \in D$,

$$\rho(z_1, z_2) = 2\tanh^{-1}\left|\frac{z_1 - z_2}{1 - \bar{z}_1 z_2}\right|.$$

Proof Lemma 5.10 (i) implies that $\rho(0, re^{i\theta}) = \rho(0, r)$. By definition,

$$\rho(0, r) = \int_0^r \frac{2dt}{1 - t^2} = 2\tanh^{-1} r.$$

In general, let l be the unique hyperbolic line through z_1 and z_2. Apply the element $\frac{z-z_1}{1-\bar{z}_1 z}$ of G; since z_1 maps to 0, we know that l goes to a diameter. By rotating, we may assume it goes to the real axis, and

$$z_2 \mapsto \left|\frac{z_2 - z_1}{1 - \bar{z}_1 z_2}\right| = r > 0.$$

Then $\rho(z_1, z_2) = \rho(0, r) = 2\tanh^{-1} r$. $\qquad \square$

For various problems (especially when there is a distinguished point, which may be sent to 0), the disc model is more convenient for calculations than the upper half-plane model. As an example, let us consider the case of hyperbolic circles, the locus of points at fixed hyperbolic distance from some point (the hyperbolic centre). We show that these are also Euclidean circles (albeit in general with different centres). As our isometry between D and H is given by a Möbius transformation, we need only prove this in the disc model (since the image of a Euclidean circle contained in D is a Euclidean circle contained in H — it cannot be a straight line, since that would intersect the boundary of H, namely $\mathbf{R} \cup \{\infty\}$). Since the above group G of Möbius transformations on D consists of isometries, the elements of G preserve both Euclidean and hyperbolic circles. The action of G is however transitive, and so we may take the hyperbolic centre of the hyperbolic circle to be at $0 \in D$; then it is clear that a hyperbolic circle of radius ρ is a Euclidean circle with radius $\tanh \frac{1}{2}\rho$.

Since isometries also preserve areas by Proposition 4.7, we can also calculate the area of a hyperbolic circle of radius ρ, by assuming that its hyperbolic centre is at $0 \in D$. The reader may then check easily that the area is given by the integral

$$2\pi \int_0^{\tanh \frac{1}{2}\rho} 4r\,dr/(1-r^2)^2 = 4\pi \left(1 - \tanh^2(\rho/2)\right)^{-1}$$

$$= 4\pi \cosh^2(\rho/2) = 2\pi(\cosh \rho - 1).$$

In passing we remark that, since open balls in the hyperbolic metric are open balls also in the Euclidean metric, the topologies (i.e. open sets) defined by the two metrics are the same. So D is homeomorphic to the Euclidean disc. In contrast to the Euclidean metric however, the hyperbolic metric may be checked to be *complete* — this basically follows from the fact that the distance from any point in D to the boundary circle is infinite in this metric.

We deduced above that any hyperbolic circle in H is a Euclidean circle. Suppose that the hyperbolic centre is at $a + ib$, with $b > 0$, and the hyperbolic radius is d; we claim that the corresponding Euclidean circle has centre at $a + ib \cosh d$, and radius $b \sinh d$. To see this, we may clearly reduce to the case when $a = 0$, by translation. The hyperbolic circle is then symmetric about the positive imaginary axis L^+, intersecting it at points iy_1 and iy_2 say, with $y_1 < y_2$. Since $\rho(ib, iy_2) = \log(y_2/b) = d$, we deduce that $y_2 = be^d$; similarly, $y_1 = be^{-d}$. The Euclidean centre is therefore at $i(y_1 + y_2)/2 = ib \cosh d$, and the Euclidean radius is $(y_2 - y_1)/2 = b \sinh d$. From this calculation, it follows that Euclidean circles in H are also hyperbolic circles.

5.4   Reflections in hyperbolic lines

We start with two basic lemmas.

Lemma 5.13   *Given a point P and a hyperbolic line $l \not\ni P$, there exists precisely one hyperbolic line $l' \ni P$ which cuts l orthogonally, say at Q, and $\rho(P, Q)$ is the minimal distance from P to l.*

Proof   We work in the disc model of the hyperbolic plane, and using the transitivity of the group G, we may take $P = 0$; the result is then obvious using Proposition 5.12, as minimizing the hyperbolic distance from 0 to l will be equivalent to minimizing the Euclidean distance.

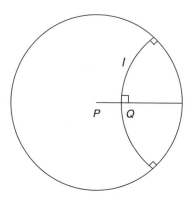

Lemma 5.14 *Suppose g is an isometry of the upper half-plane model, fixing all points of L^+; then either $g = id$, or $g(z) = -\bar{z}$ for all $z \in H$ (the latter being a reflection in the y-axis, which is an isometry, since $|dz|^2/y^2$ is clearly invariant).*

Proof For $P \notin L^+$, there exists a unique hyperbolic line l' through P and perpendicular to L^+ (l' is a semicircle with centre 0).

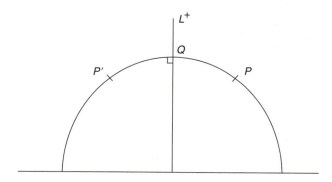

Since g is an isometry, the minimum distance from $g(P)$ to L^+ is the same as the minimum distance from P to L^+, and moreover this distance is $\rho(P, Q) = \rho(g(P), Q)$. As l' is the unique hyperbolic line passing through Q and perpendicular to L^+, it follows (using Lemma 5.13 again) that $g(P) \in l'$. Since $\rho(P, Q) = \rho(g(P), Q)$, we deduce that $g(P) = P$ or P' (where P' denotes the image of P under the reflection). The lemma is then implied by the following claim.

Claim If there exists $P \notin L^+$ such that $g(P) = P$, then $g = id$; otherwise g is the reflection in L^+.

To prove the claim, we may assume (by symmetry) that $P \in H^+ = \{z \in H : \mathrm{Re}\, z > 0\}$. Let A be any point of H^+, and construct hyperbolic lines as shown.

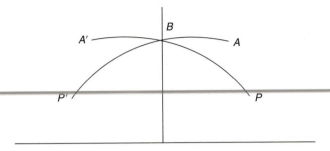

If $g(A) = A'$, then $\rho(A', P) = \rho(A, P)$. But

$$\rho(A', P) = \rho(A', B) + \rho(B, P)$$
$$= \rho(A, B) + \rho(B, P).$$

The triangle inequality then implies that B is on the hyperbolic line segment PA, and this then contradicts the fact that both P and A are in H^+. It therefore follows that $g(A) = A$ for all points $A \in H$.

The alternative is that every point not on L^+ is reflected.   □

Definition 5.15   Let R denote the reflection in the y-axis; for any hyperbolic line l, we choose $T \in PSL(2, \mathbf{R})$ such that $T(l) = L^+$. Then $R_l = T^{-1}RT \neq$ id is an isometry of H which fixes every point of l, and by Lemma 5.14 it is uniquely defined by this property. We call this isometry the *reflection* in the hyperbolic line l.

Clearly R_l may also be defined geometrically; for $P \in H \setminus l$, we can drop the perpendicular hyperbolic line l' to l by Lemma 5.13, and then let $R_l(P) \neq P$ be the obvious point on l', whose distance from l is the same as that of P.

Proposition 5.16   *Any isometry g of H is either an element of $PSL(2, \mathbf{R})$, or else an element of the coset $PSL(2, \mathbf{R})R$.*

Proof   Suppose $g(L^+) = l$. Choose $T \in PSL(2, \mathbf{R})$ such that $Tl = L^+$ and consider Tg instead of g, which is therefore an isometry sending L^+ to itself.

In this way, we reduce to the case when g sends L^+ to itself, and, composing if necessary with $z \mapsto -\frac{1}{z}$, we may assume that $g(0) = 0$, $g(\infty) = \infty$. Scaling by a real number, we may also assume $g(i) = i$. From this, it is clear that g (being an isometry) fixes all points of L^+.

By Lemma 5.14, we deduce that $g =$ id or R. Hence the result follows.   □

Remark 5.17   Therefore, $PSL(2, \mathbf{R})$ is an index two subgroup of the full isometry group. The isometries of H are all of the form

$$z \mapsto \frac{az + b}{cz + d} \qquad \text{or} \qquad z \mapsto \frac{a(-\bar{z}) + b}{c(-\bar{z}) + d},$$

with a, b, c, d real and $ad - bc = 1$. Those in $PSL(2, \mathbf{R})$ are called the *direct isometries*.

Recall that, for both the Euclidean plane and the sphere, any isometry could be expressed as a product of at most three reflections. The same result, with an analogous proof, holds for the hyperbolic plane. Firstly, we need a lemma about perpendicular bisectors of hyperbolic line segments.

Lemma 5.18   *Given two points P and Q in the hyperbolic plane, the locus of points equidistant from P and Q is a hyperbolic line l, the perpendicular bisector of the hyperbolic line segment from P to Q. In particular, the reflection R_l is an isometry, fixing all points of l but swapping the points P and Q.*

Proof   We shall work in the disc model. There is a unique hyperbolic line l' through P and Q; we let M be the mid-point of the hyperbolic line segment PQ, and then apply an isometry from the group G which sends M to the origin and l' to the real diameter.

We may assume therefore that we are in the situation illustrated below. There is then a unique hyperbolic line l through 0 which cuts the hyperbolic line segment PQ at right-angles, namely the imaginary diameter of D. By symmetry the points of l are equidistant from P and Q.

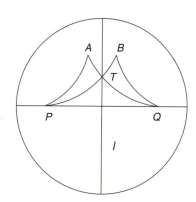

We just need to show now that any point A which is equidistant from P and Q lies on l. Suppose there is such a point A not lying on the imaginary diameter; by symmetry there is a second such point B obtained by reflection of A in l. A contradiction will now occur in exactly the same way as in the proof of Lemma 5.14. By symmetry, the hyperbolic line segments PB and QA intersect l in a common point T say. Using symmetry again gives

$$d(P,A) = d(Q,A) = d(P,B) = d(P,T) + d(T,B) = d(P,T) + d(T,A).$$

The triangle inequality then implies that T is on the hyperbolic line segment PA, and this then yields a contradiction (as in the proof of Lemma 5.14).

The last sentence is now clear.   □

Proposition 5.19   *Any isometry g of the hyperbolic plane may be expressed as the product of at most three reflections. The direct isometries are those which are the product of two reflections.*

Proof   Now we have Lemma 5.18, the proof follows in exactly the same way as the Euclidean case. Consider say the three points $i, 2i$ and $1 + i$ in H. Using the previous lemma and arguing exactly as in Theorem 1.7, we can find an isometry h, which is the product of at most three reflections, such that $h \circ g$ fixes $i, 2i$ and $1 + i$ — recall that the first step is to compose g (if necessary) with the reflection which swaps the points i and $g(i)$. However, an isometry which fixes i and $2i$ must fix all points of L^+, and if it also fixes $1 + i$, it must be the identity by Lemma 5.14. Therefore $g = h^{-1}$ is the product of at most three reflections.

The second sentence follows from Remark 5.17.   □

5.5   Hyperbolic triangles

Definition 5.20   A hyperbolic triangle ABC is defined by three hyperbolic line segments, as illustrated.

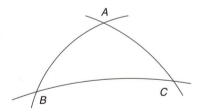

We comment that hyperbolic triangles are *convex*, meaning that the hyperbolic line segment joining any two points of the triangle lies entirely in the triangle (Exercise 5.8). As a degenerate case, we shall include the cases where some of the vertices are on the boundary of the hyperbolic plane (for the upper half-plane model, on **R** or at ∞; for the disc model, on the unit circle).

In the upper half-plane model, the area of a region $R \subset H$ is by definition

$$\text{area } R = \iint_R \frac{dx\,dy}{y^2}.$$

Locally, we are just scaling Euclidean lengths by a factor of $1/y$, and Euclidean areas by a factor of $1/y^2$.

Theorem 5.21 (Gauss–Bonnet) *For a hyperbolic triangle $T = ABC$ with angles α, β, γ (possibly with some angles zero),*

$$\text{area } T = \pi - (\alpha + \beta + \gamma).$$

Proof   Recall first that areas are invariant under isometries, and so we may choose either the disc or upper half-plane model according to convenience, and may also apply any isometry we wish of the model in order for the hyperbolic triangle to be conveniently placed. We choose to work in H.

We prove this result first under the assumption $\gamma = 0$. Take T in the upper half-plane model, where we may take C at ∞. Translating and scaling by a real number, we may assume also that A, B are on the circle $x^2 + y^2 = 1$ (including the cases where one or both are on the real axis).

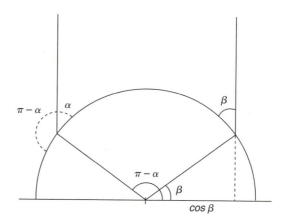

Therefore,

$$\text{area } T = \int_{\cos(\pi-\alpha)}^{\cos \beta} \int_{(1-x^2)^{1/2}}^{\infty} \frac{dy}{y^2}\, dx$$

$$= \int_{\cos(\pi-\alpha)}^{\cos \beta} \frac{dx}{(1-x^2)^{1/2}}$$

$$= [-\cos^{-1} x]_{\cos(\pi-\alpha)}^{\cos \beta}$$

$$= \pi - \alpha - \beta.$$

In general, we can express any triangle as the difference of two triangles with a common vertex at ∞ — working in the upper half-plane model, this is easy to see.

We can always arrange that one side of a triangle ABC is a vertical line.

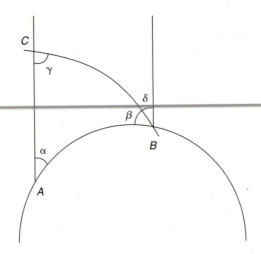

The diagram shows two triangles with a vertex at ∞, namely $\Delta_1 = AB\infty$ and $\Delta_2 = CB\infty$. From the previous case, we know that

$$\text{area } \Delta_1 = \pi - \alpha - (\beta + \delta)$$
$$\text{area } \Delta_2 = \pi - \delta - (\pi - \gamma).$$

Subtracting these two expressions, we obtain

$$\text{area } \Delta = \pi - (\alpha + \beta + \gamma). \qquad \square$$

Corollary 5.22   *The area of a hyperbolic n-gon (sides being hyperbolic line segments) is given by the formula*

$$(n - 2)\pi - (\alpha_1 + \cdots + \alpha_n),$$

where $\alpha_1, \ldots, \alpha_n$ are the internal angles.

Proof   This follows from Theorem 5.21, precisely as in the proof of Proposition 2.16, where we obtained the area of a spherical polygon contained in a hemisphere by an inductive combinatorial argument. The fact that there is a locally convex vertex follows in the same way as before. The only other property of the hemisphere that we used was its convexity, that through any two points of the hemisphere there is a unique spherical line segment of minimal length joining them, and this spherical line segment is contained in the hemisphere. In the hyperbolic plane, there is always a unique hyperbolic line segment of minimum length through any two points, and so the whole hyperbolic plane is convex. Moreover, we commented above that hyperbolic triangles are also convex. The inductive argument therefore goes through unchanged. $\square$

Proposition 5.23 *For $n \geq 3$, given any α with $0 < \alpha < (1 - \frac{2}{n})\pi$, there is a regular hyperbolic n-gon, all of whose angles are α.*

Proof This follows from a rather elegant continuity argument. We consider the points $re^{2k\pi i/n}$, for $0 < r \leq 1$ and $k = 1, \ldots, n$, in the unit disc. For a given r, these determine a regular hyperbolic n-gon in D; let $\alpha(r)$ denote the value of any interior angle of this hyperbolic n-gon. Recall that the area of the n-gon is $(n - 2)\pi - n\alpha(r)$; since the area varies continuously with r, we deduce that $\alpha(r)$ is a continuous monotonic function of r. We can take $r = 1$, and the polygon will have all its vertices on the boundary. As the sides are orthogonal to the boundary, the angle between any two adjacent sides is then zero; thus $\alpha(r) \to 0$ as $r \to 1$. If we now consider what happens as $r \to 0$, the area will clearly tend to zero, and hence $\alpha(r) \to (1 - \frac{2}{n})\pi$ as $r \to 0$. The Intermediate Value theorem for continuous functions of one variable therefore implies the result. $\qquad \square$

In particular, for any integer $g \geq 2$, there is a regular hyperbolic $4g$-gon whose angles are all $\pi/2g$. We may however make the identifications of sides as described in Chapter 3 in order to obtain topologically a g-holed torus. In this case, there is no longer an obstruction to extending the hyperbolic metric on the $4g$-gon to a locally hyperbolic metric on the g-holed torus, since we have arranged that the angles of our regular *hyperbolic* $4g$-gon do add up to 2π; in fact one can argue along these lines to prove that there does exist such a locally hyperbolic metric on the g-holed torus.

5.6 Parallel and ultraparallel lines

Two distinct spherical lines in S^2 meet (in two points); two distinct lines in $\mathbf{P}^2(\mathbf{R})$ meet in one point. Two distinct lines in $\mathbf{R}^2$ meet (in one point) if and only if they are not parallel.

Definition 5.24 Taking say the disc model for the hyperbolic plane, two hyperbolic lines l_1 and l_2 in D are said to be *parallel* if they meet only at the boundary $|z| = 1$. They are called *ultraparallel* if they do not meet anywhere in the closed disc $|z| \leq 1$.

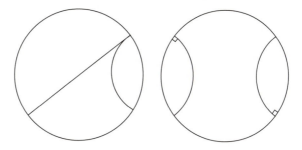

Remark In the Euclidean plane, the parallel axiom may be stated as follows: Given a line l and a point $P \notin l$, there exists a unique line $l' \ni P$ with $l \cap l' = \emptyset$. We have seen that this fails for both spherical and hyperbolic geometry, but for very different reasons. For the hyperbolic plane, it fails since l' is not unique.

Definition 5.25   Given any two subsets A, B of a metric space (X, ρ), we define the *distance* between the sets to be

$$d(A, B) = \inf\{\rho(P, Q) : P \in A, Q \in B\}.$$

We now show that ultraparallel lines are characterized by the distance between them being non-zero (a slightly simpler proof of this fact may be found in Exercise 5.11).

Proposition 5.26   *Suppose that l_1 and l_2 are disjoint hyperbolic lines in the hyperbolic plane. If l_1 and l_2 are parallel, then $d(l_1, l_2) = 0$; if l_1 and l_2 are ultraparallel, then $d(l_1, l_2) > 0$.*

Proof   We work in the upper half-plane model, and take l_1 to be the positive imaginary axis. If l_2 is parallel to l_1, we may take it to be a vertical half-line given by $x = a > 0$. For any $b > 0$, the distance $\rho(ib, a + ib)$ is at most the length of the straight line segment between the two points given by $y = b$, namely a/b (since Euclidean distances are locally scaled by a factor $1/y$). As $b \to \infty$, we deduce that $\rho(ib, a + ib) \to 0$, and hence $d(l_1, l_2) = 0$.

If however l_2 is ultraparallel to l_1; we may assume that l_2 is a semicircle with centre $a > 0$ on the real line, and radius r, where $0 < r < a$. For any point $P \in l_2$, we obtain $d(P, l_1)$ as the length of the hyperbolic line segment resulting from dropping the perpendicular from P to l_1 (Lemma 5.13). Since the hyperbolic lines orthogonal to l_1 are just semicircles with centre at 0, we have that $d(P, l_1)$ is the length of the segment (between P and l_1) of the semicircle with centre at 0 which passes through P. The radius of this semicircle is at most $a + r$.

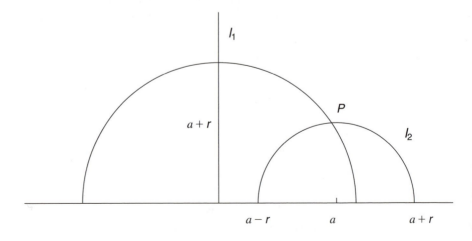

We denote this curve by $\gamma = (\gamma_1, \gamma_2) : [0,1] \to H$. So

$$d(P, l_1) = \int_0^1 \left((\gamma_1'(t))^2 + (\gamma_2'(t))^2 \right)^{1/2} dt/\gamma_2(t)$$

$$\geq \int_0^1 |\gamma_1'(t)| \, dt/(a+r) \geq \int_0^1 \gamma_1'(t) dt/(a+r) \geq (a-r)/(a+r),$$

since the difference in the x-coordinate is at least $a - r$. This bound is however independent of the point P on l_2, and hence

$$d(l_1, l_2) \geq (a-r)/(a+r) > 0.$$

$\square$

5.7 Hyperboloid model of the hyperbolic plane

There is another basic model of the hyperbolic plane which should be mentioned, namely the *hyperboloid model*. This model is entirely analogous to our study of the sphere as an embedded surface in $\mathbf{R}^3$, but with a few signs in inner-products changed. One advantage of this model is that one can prove formulae which occur in hyperbolic trigonometry with a minimum of calculation, much as we did when proving the spherical versions. If one wants to prove these formulae without using the hyperboloid model, it will involve the manipulation of formulae — a good place to start would be the formula in Exercise 5.7.

Let us take a Lorentzian inner-product $\langle\!\langle \, , \, \rangle\!\rangle$ on $\mathbf{R}^3$, corresponding to the matrix

$$\begin{pmatrix} 1 & 0 & 0 \\ 0 & 1 & 0 \\ 0 & 0 & -1 \end{pmatrix}.$$

Set $q(\mathbf{x}) = \langle\!\langle \mathbf{x}, \mathbf{x} \rangle\!\rangle = x^2 + y^2 - z^2$ and let S be the surface $q(\mathbf{x}) = -1$, the hyperboloid of two sheets $x^2 + y^2 = z^2 - 1$. Let S^+ be the upper-sheet of the hyperboloid (i.e. $z > 0$). Now consider the stereographic projection map π from the point $(0, 0, -1)$, mapping S^+ onto the unit disc D, given by

$$\pi(x, y, z) = \frac{x + iy}{1 + z}.$$

Then

$$u + iv = \frac{x + iy}{1 + z}$$

$$\implies u^2 + v^2 = \frac{1 - z^2}{(1 + z)^2} = \frac{1 - z}{1 + z}.$$

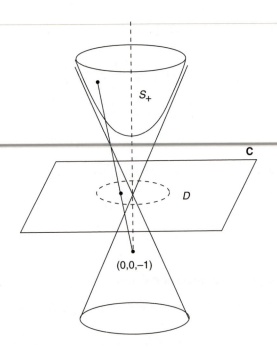

So if $r = u^2 + v^2$,

$$z = \frac{1 + r^2}{1 - r^2} \quad \Longrightarrow \quad 1 + z = \frac{2}{1 - r^2}.$$

Therefore

$$x = (1 + z)u = \frac{2u}{1 - r^2}, \qquad y = \frac{2v}{1 - r^2}.$$

Now

$$\frac{\partial x}{\partial u} = \frac{\partial}{\partial u}\left(\frac{2u}{1 - u^2 - v^2}\right) = \frac{2}{(1 - r^2)^2}(1 + u^2 - v^2),$$

$$\frac{\partial y}{\partial u} = \frac{4uv}{(1 - r^2)^2},$$

$$\frac{\partial z}{\partial u} = \frac{\partial(1 + z)}{\partial u} = \frac{4u}{(1 - r^2)^2}.$$

Setting

$$\sigma(u, v) = \left(\frac{2u}{1 - r^2}, \frac{2v}{1 - r^2}, \frac{1 + r^2}{1 - r^2}\right),$$

we have

$$\sigma_u := \frac{\partial \sigma}{\partial u} = \frac{2}{(1-r^2)^2}\left(1+u^2-v^2, 2uv, 2u\right),$$

$$\sigma_v := \frac{\partial \sigma}{\partial v} = \frac{2}{(1-r^2)^2}\left(2uv, 1+v^2-u^2, 2v\right),$$

the second equation following from the first by symmetry.

The *tangent space* to S^+ at a point $\mathbf{a} \in S^+$ will by definition consist of vectors $\mathbf{x}$ such that, for t small, $\langle\langle \mathbf{a}+t\mathbf{x}, \mathbf{a}+t\mathbf{x}\rangle\rangle = -1 + O(t^2)$, i.e. with $\langle\langle \mathbf{a}, \mathbf{x}\rangle\rangle = 0$. However, with σ as above, we have $\langle\langle \sigma(u,v), \sigma(u,v)\rangle\rangle = -1$, and so, differentiating with respect to u and v, we see that both σ_u and σ_v are in the tangent space at $\sigma(u,v)$. A routine check verifies that they are always linearly independent for $u^2 + v^2 < 1$, and so σ_u, σ_v form a basis for the tangent space to S^+ at $\sigma(u,v)$. Moreover, $\langle\langle\ ,\ \rangle\rangle$ determines a symmetric bilinear form on this vector space, and hence via $d\sigma$ (which identifies e_1 with σ_u and e_2 with σ_v) corresponds to a symmetric bilinear form on $\mathbf{R}^2$. This bilinear form may be written as $E\,du^2 + 2F\,du\,dv + G\,dv^2$, where

$$E = \langle\langle\sigma_u, \sigma_u\rangle\rangle = \frac{4}{(1-r^2)^4}\left(\left(1+u^2-v^2\right)^2 + 4u^2v^2 - 4u^2\right)$$

$$= \frac{4}{(1-r^2)^4}\left(1-\left(u^2+v^2\right)\right)^2 = \frac{4}{(1-r^2)^2},$$

$$F = \langle\langle\sigma_u, \sigma_v\rangle\rangle = 0 \quad \text{by inspection, and}$$

$$G = \langle\langle\sigma_v, \sigma_v\rangle\rangle = \frac{4}{(1-r^2)^2} \quad \text{by symmetry.}$$

This then is just the hyperbolic metric on the Poincaré disc model of the hyperbolic plane.

We now look for isometries of the hyperboloid model. Let $O(2, 1)$ denote the group of 3×3 matrices which preserve the above Lorentzian inner-product; these are just the matrices P for which

$$P^t \begin{pmatrix} 1 & 0 & 0 \\ 0 & 1 & 0 \\ 0 & 0 & -1 \end{pmatrix} P = \begin{pmatrix} 1 & 0 & 0 \\ 0 & 1 & 0 \\ 0 & 0 & -1 \end{pmatrix}.$$

In particular, the action of $O(2, 1)$ on $\mathbf{R}^3$ preserves the hyperboloid. We however want the elements which send the upper-sheet S^+ of the hyperboloid to itself and do not switch the sheets. This further condition defines an index two subgroup $O^+(2, 1)$ of $O(2, 1)$. Note that

$$\begin{pmatrix} 1 & 0 & 0 \\ 0 & 1 & 0 \\ 0 & 0 & -1 \end{pmatrix}$$

is in $O(2, 1)$ but not in $O^+(2, 1)$. If P is in $O(2, 1)$, then by taking determinants of the defining equation, we deduce that $\det(P)^2 = 1$, that is $\det(P) = \pm 1$.

Since the inner-products on the tangent spaces are determined by the Loretzian inner-product on $\mathbf{R}^3$, we deduce furthermore that any element of $O^+(2, 1)$ will preserve the metric. It will moreover preserve the lengths of curves on S^+ in the given metric, where for a given curve $\gamma(t) = (\gamma_1(t), \gamma_2(t), \gamma_3(t))$ with $\gamma : [0, 1] \to S^+$, its length is defined to be

$$\int_0^1 \langle\langle \dot\gamma(t), \dot\gamma(t) \rangle\rangle \, dt = \int_0^1 (\dot\gamma_1^2 + \dot\gamma_2^2 - \dot\gamma_3^2)^{1/2} dt.$$

There are two types of element of $O^+(2, 1)$ which will be of particular interest to us. The first are just the rotations about the z-axis, that is matrices of the following form:

$$\begin{pmatrix} \cos\theta & -\sin\theta & 0 \\ \sin\theta & \cos\theta & 0 \\ 0 & 0 & 1 \end{pmatrix}$$

for $0 \le \theta < 2\pi$. This matrix is clearly in $O(2, 1)$, and, as any such matrix may be joined to the identity matrix I by a continuous curve of such matrices in $O(2, 1)$, the matrix is in $O^+(2, 1)$.

The second type of element which is of interest to us is those of the form:

$$\begin{pmatrix} 1 & 0 & 0 \\ 0 & \cosh d & -\sinh d \\ 0 & -\sinh d & \cosh d \end{pmatrix}$$

for $d \ge 0$. Again, this matrix is readily checked to be in $O(2, 1)$, and as any such matrix may be joined to the identity matrix I by a continuous curve of such matrices in $O(2, 1)$, the matrix is in $O^+(2, 1)$. This matrix P has the useful property that

$$P \begin{pmatrix} 0 \\ \sinh d \\ \cosh d \end{pmatrix} = \begin{pmatrix} 0 \\ 0 \\ 1 \end{pmatrix}.$$

With the aid of elements of the above two types, we see that the action of $O^+(2, 1)$ on S^+ is transitive. Given an arbitrary vector $(x, y, z)^t \in S^+$, we can first rotate it to a vector of the form $(0, \sinh d, \cosh d)^t$ by means of an element of the first type, and then send this to to $(0, 0, 1)^t$ by means of an element of the second type.

We note, for later use, that matrices of the above two types have determinant $+1$; one can in fact show that elements of these two types generate $O^+(2, 1)$ (cf. the proof of Theorem 2.19), and so $\det(P) = 1$ for all $P \in O^+(2, 1)$ (we shall not however need this latter fact).

Given two arbitrary points, with the aid of elements of the above two types, we may send the first point to $(0, 0, 1)$, and the second point to $(0, \sinh d, \cosh d)$ for some $d > 0$. Under the stereographic projection map, these correspond to the points 0 and $i \sinh d / (1 + \cosh d) = i \tanh(d/2)$ in the disc model D. Since distances on S^+ correspond to distances in D, we see that the two points are distance d apart. Moreover, we note that the Lorentzian inner product of the two vectors is $- \cosh d$. We deduce therefore that for any two vectors $\mathbf{x}$ and $\mathbf{y}$ representing points on S^+, we have $\langle\langle \mathbf{x}, \mathbf{y} \rangle\rangle = - \cosh d$, where d is the hyperbolic distance between the two points.

We are now in a position to prove the hyperbolic cosine formula. It is possible to prove this in exactly the same way as we proved the spherical cosine formula, by defining an appropriate cross-product of vectors in $\mathbf{R}^3$ associated to the Lorentzian inner-product defined above, but it is probably clearer to use the above isometries.

Proposition 5.27 (Hyperbolic cosine formula) *Let $\triangle$ be a hyperbolic triangle, with angles α, β, γ, and sides of length a, b, c (the side of length a being opposite the vertex with angle α, and similarly for b and c). Then*

$$\cosh a = \cosh b \, \cosh c \, - \, \sinh b \, \sinh c \, \cos \alpha.$$

Proof   We work on the hyperboloid model, where we may assume that the vertices of the triangle correspond to points $A = (0, 0, 1)$, $B = (0, \sinh c, \cosh c)$, and

$$C = (\sin \alpha \, \sinh b, \, \cos \alpha \, \sinh b, \, \cosh b)$$

on S^+. If the metric on S^+ is denoted by d, then $d(A, B) = c$, $d(A, C) = b$ and $d(B, C) = a$. From the above discussion, $- \cosh a$ is just the Lorentzian inner-product of the position vectors $\mathbf{B}$ and $\mathbf{C}$, namely

$$\cos \alpha \, \sinh b \, \sinh c - \cosh b \, \cosh c.$$

$\square$

Proposition 5.28 (Hyperbolic sine formula) *With the notation as above,*

$$\sinh a / \sin \alpha = \sinh b / \sin \beta = \sinh c / \sin \gamma.$$

Proof   Given points A, B, C, we can consider the matrix $M (A, B, C)$ whose columns are the position vectors $\mathbf{A}$, $\mathbf{B}$ and $\mathbf{C}$, and take its determinant. Clearly, this is invariant when we operate on $\mathbf{R}^3$ by elements of the above two types, as the determinant of such matrices was seen to be one. We may therefore reduce down to the points A, B, C being in the particularly simple form assumed in the proof of the previous result. With these three points, the determinant of $M (A, B, C)$ is checked to be $- \sinh c \, \sinh b \, \sin \alpha$, and this was therefore the number obtained from our original three points. The determinant of the matrix is however invariant under cyclic permutations of the points, from which we deduce

$$\sinh c \, \sinh b \, \sin \alpha = \sinh a \, \sinh c \, \sin \beta = \sinh b \, \sinh a \, \sin \gamma.$$

Dividing through by $\sinh a \, \sinh b \, \sinh c$ gives the result required. $\square$

Exercises

5.1   Verify that the Möbius transformation $\zeta \mapsto i(1+\zeta)/(1-\zeta)$ from the disc model of the hyperbolic plane to the upper half-plane model may be defined by an element of $SU(2)$.

5.2   Suppose that z_1, z_2 are points in the upper half-plane, and suppose the hyperbolic line through z_1 and z_2 meets the real axis at points z_1^* and z_2^*, where z_1 lies on the hyperbolic line segment $[z_1^*, z_2]$, and where one of z_1^* and z_2^* might be ∞. Show that the hyperbolic distance $\rho(z_1, z_2) = \log r$, where r is the cross-ratio of the four points z_1^*, z_1, z_2, z_2^*, taken in an appropriate order.

5.3   If a is a point of the upper half-plane, show that the Möbius transformation g given by

$$g(z) = \frac{z-a}{z-\bar{a}}$$

defines an isometry from the upper half-plane model H to the disc model D of the hyperbolic plane, sending a to zero. Deduce that for points z_1, z_2 in the upper half-plane, the hyperbolic distance is given by $\rho(z_1, z_2) = 2\tanh^{-1}\left|\frac{z_1-z_2}{z_1-\bar{z_2}}\right|$.

5.4   Let l denote the hyperbolic line in H given by a semicircle with centre $a \in \mathbf{R}$ and radius $r > 0$. Show that the reflection R_l is given by the formula

$$R_l(z) = a + \frac{r^2}{\bar{z}-a}.$$

5.5   Given two hyperbolic lines meeting at a point, show that the locus of points equidistant from the two lines forms two further hyperbolic lines through the point. Show that in a hyperbolic triangle, none of whose vertices are at infinity, the angle bisectors are concurrent.

5.6   Show that any isometry g of the disc model D for the hyperbolic plane is *either* of the form (for some $a \in D$ and $0 \le \theta < 2\pi$)

$$g(z) = e^{i\theta}\frac{z-a}{1-\bar{a}z},$$

or of the form

$$g(z) = e^{i\theta}\frac{\bar{z}-a}{1-\bar{a}\bar{z}}.$$

5.7   For arbitrary points z, w in $\mathbf{C}$, prove the identity

$$|1-\bar{z}w|^2 = |z-w|^2 + (1-|z|^2)(1-|w|^2).$$

Given points z, w in the unit disc model of the hyperbolic plane, prove the identity

$$\sinh^2(\rho(z, w)/2) = \frac{|z - w|^2}{(1 - |z|^2)(1 - |w|^2)},$$

where ρ denotes the hyperbolic distance.

5.8 Let A be the sector in the unit disc model of the hyperbolic plane given by $0 \le \theta \le \alpha$ for some $\alpha < \pi$. Show that A is *convex*, i.e. for any points $P, Q \in A$, the hyperbolic line segment PQ lies entirely in A. If B is a vertical strip in the upper half-plane model of the hyperbolic plane, show that B is convex. Deduce that any hyperbolic triangle is the intersection of three suitable convex subsets of the hyperbolic plane, and is therefore itself a convex subset.

5.9 Let T be a hyperbolic triangle; show that the radius of any inscribed hyperbolic circle is less than $\cosh^{-1}(3/2)$. Generalize this result to hyperbolic polygons.

5.10 If α and β are positive numbers with $\alpha + \beta < \pi$, show that there exists a hyperbolic triangle (one vertex at infinity) with angles 0, α and β. For any positive numbers α, β and γ, with $\alpha + \beta + \gamma < \pi$, show that there exists a hyperbolic triangle with these angles. [Hint: For the second part, you may need a continuity argument.]

5.11 Show that two distinct hyperbolic lines have a common perpendicular if and only if they are ultraparallel, and that in this case the perpendicular is unique. By taking this common perpendicular to be the imaginary axis, find a simple proof for Proposition 5.26.

5.12 Show that the composite of two reflections in distinct hyperbolic lines has finite order if and only if the lines meet at a point in (the interior of) the hyperbolic plane with angle which is a rational multiple of π. [Hint: For the ultraparallel case, use the argument from the previous exercise.]

5.13 Fix a point P on the boundary of D, the disc model of the hyperbolic plane. Give a description of the curves in D that are orthogonal to every hyperbolic line that passes through P.

5.14 If two hyperbolic triangles have the same side-lengths, including the degenerate cases of some sides being infinite, prove that the triangles are congruent, i.e. there is an isometry of the hyperbolic plane sending one onto the other. Using the argument from Exercise 5.10, or otherwise, prove that the same holds if they have the same angles (also including the degenerate cases of some angles being zero).

5.15 If ABC is a hyperbolic triangle, with the angle at A at least $\pi/2$, show that the side BC has maximum length. Given points z_1, z_2 in the hyperbolic plane, let w be any point of the hyperbolic line segment joining z_1 to z_2, and w' be any point not on the hyperbolic line passing through the other three. Show that

$$\rho(w', w) \le \max\{\rho(w', z_1), \rho(w', z_2)\}.$$

Deduce that the *diameter* of a hyperbolic triangle $\triangle$ (that is, $\sup\{\rho(P, Q) : P, Q \in \triangle\}$) is equal to the length of its longest side. Show that the corresponding result holds for Euclidean triangles, but does not hold for spherical triangles.

5.16   Given two distinct hyperbolic circles of radius ρ, show that there is a hyperbolic circle of radius strictly less than ρ which contains their intersection. (Hint: After applying an appropriate isometry, one may assume that the hyperbolic circles have centres $-a + ib$ and $a + ib$ in H, for $a, b > 0$.)

Deduce that, for any finite set of points in the hyperbolic plane, there is a unique hyperbolic circle of minimum radius which encloses them (some will of course lie on the circle). Let G be a finite subgroup of $PSL(2, \mathbf{R})$; show that the action of G on the upper half-plane has a fixed point, and deduce that G is a cyclic group. Prove that any finite subgroup of $\mathrm{Isom}(H)$ is either cyclic or dihedral.

6 Smooth embedded surfaces

We now move on from the classical geometries, as described in previous chapters, to rather more general two-dimensional geometries. In this chapter, we study the concrete case of smooth surfaces embedded in $\mathbf{R}^3$; any such surface has a metric, defined in terms of infima of lengths of curves on the embedded surface, as was the case for the sphere. In Chapter 8, we generalize further and introduce the concept of abstract smooth surfaces, equipped with a Riemannian metric. These will represent a common generalization of both the embedded surfaces (as studied in this chapter), and general Riemannian metrics on open subsets of $\mathbf{R}^2$ (as studied in Chapter 4). In the remainder of this book, we shall develop the theory of curvature and geodesics in this more general context, thus providing an introduction to two central concepts from elementary differential geometry, and culminating in a proof of the Gauss–Bonnet theorem for arbitrary compact smooth surfaces.

6.1 Smooth parametrizations

Definition 6.1   A subset $S \subset \mathbf{R}^3$ is called a (parametrized) *smooth embedded surface* if each point of S has an open neighbourhood $U = W \cap S$ (where W is open in $\mathbf{R}^3$) and a map $\sigma : V \to U \subset S \subset \mathbf{R}^3$ from an open subset V of $\mathbf{R}^2$, such that

- $\sigma : V \to U$ is a *homeomorphism* (i.e. a continuous map with a continuous inverse),
- $\sigma(u, v) = (x(u, v), y(u, v), z(u, v))$ is smooth (i.e. it has partial derivatives of all orders),
- at each point $Q = \sigma(P)$, the vectors $\sigma_u(P) = d\sigma_P(e_1)$ and $\sigma_v(P) = d\sigma_P(e_2)$ are *linearly independent* (where $d\sigma_P : \mathbf{R}^2 \to \mathbf{R}^3$ denotes the derivative at P). Recall that

$$\sigma_u(P) = \frac{\partial \sigma}{\partial u}(P) = \begin{pmatrix} \frac{\partial x}{\partial u}(P) \\ \frac{\partial y}{\partial u}(P) \\ \frac{\partial z}{\partial u}(P) \end{pmatrix}.$$

We call (u, v) smooth coordinates on U and the subspace of $\mathbf{R}^3$ spanned by σ_u, σ_v the *tangent space* $T_{S,Q}$ to S at Q. The map σ is called a *smooth parametrization* of $U \subset S$.

Proposition 6.2   *Suppose $\sigma : V \to U$ and $\tilde{\sigma} : \tilde{V} \to U$ are smooth parametrizations of U. Then the homeomorphism $\phi = \sigma^{-1} \circ \tilde{\sigma}$ is a diffeomorphism (i.e. $\phi : \tilde{V} \to V$ and its inverse are both smooth) with $\tilde{\sigma} = \sigma \circ \phi$.*

Proof   The Jacobian matrix for $\sigma(u, v) = (x(u, v), y(u, v), z(u, v))$,

$$\begin{pmatrix} x_u & x_v \\ y_u & y_v \\ z_u & z_v \end{pmatrix},$$

has rank 2 everywhere on V. Since ϕ clearly is a homeomorphism, we only need to show that it is a diffeomorphism locally. Without loss of generality, we assume $\det\begin{pmatrix} x_u & x_v \\ y_u & y_v \end{pmatrix} \neq 0$ at $(u_0, v_0) \in V$; we consider $F : V \to \mathbf{R}^2$ given by composing σ with the projection map $\mathbf{R}^3 \to \mathbf{R}^2$ onto the first two coordinates, namely $F(u, v) = (x(u, v), y(u, v))$. Clearly F is a smooth function.

Since the Jacobian matrix of F is non-singular at (u_0, v_0), the Inverse Function theorem (see Section 4.1) implies that F is a local diffeomorphism at (u_0, v_0). Hence, there exist open neighbourhoods $(u_0, v_0) \in N \subset V$ and $F(u_0, v_0) \in N' \subset \mathbf{R}^2$, such that $F|_N : N \to N'$ is a diffeomorphism. Since $\sigma|_N : N \to \sigma(N)$ is a homeomorphism onto an open subset of U, and $F|_N : N \to N'$ is also a homeomorphism, so too is the projection map $\pi : \sigma(N) \to N' \subset \mathbf{R}^2$. Now $\tilde{N} := \tilde{\sigma}^{-1}(\sigma(N))$ is open in $\tilde{V}$ and $\sigma^{-1} \circ \tilde{\sigma} = \sigma^{-1} \circ \pi^{-1} \circ \pi \circ \tilde{\sigma} = F^{-1} \circ \tilde{F}$ on $\tilde{N}$, where $\tilde{F} = \pi \circ \tilde{\sigma}$.

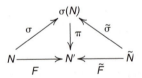

Both $\tilde{F}|_{\tilde{N}}$ and $F^{-1}|_{N'}$ are smooth maps, and therefore so too is the composite $\phi|_{\tilde{N}}$. Hence, by symmetry, both $\phi : \tilde{V} \to V$ and its inverse are smooth.   □

Corollary 6.3   *The tangent space $T_{S,Q}$ is independent of the parametrization $\sigma : V \to U \ni Q$.*

Proof   By Proposition 6.2, the smooth parametrizations of U are precisely of the form $\tilde{\sigma} = \sigma \circ \phi$ with $\phi = (\phi_1, \phi_2) : \tilde{V} \to V \subset \mathbf{R}^2$ a diffeomorphism from the open subset $\tilde{V} \subset \mathbf{R}^2$ to V. Suppose the coordinates on $\tilde{V}$ are $(\tilde{u}, \tilde{v})$. By the Chain Rule,

$$\tilde{\sigma}_{\tilde{u}} = \frac{\partial \phi_1}{\partial \tilde{u}} \sigma_u + \frac{\partial \phi_2}{\partial \tilde{u}} \sigma_v,$$

$$\tilde{\sigma}_{\tilde{v}} = \frac{\partial \phi_1}{\partial \tilde{v}} \sigma_u + \frac{\partial \phi_2}{\partial \tilde{v}} \sigma_v.$$

We set

$$J(\phi) := \begin{pmatrix} \partial\phi_1/\partial\tilde{u} & \partial\phi_1/\partial\tilde{v} \\ \partial\phi_2/\partial\tilde{u} & \partial\phi_2/\partial\tilde{v} \end{pmatrix}$$

to be the Jacobian matrix for ϕ (i.e. it represents $d\phi$ with respect to the standard basis e_1, e_2). Since $J(\phi)$ is invertible, the subspace of $\mathbf{R}^3$ spanned by $\tilde{\sigma}_{\tilde{u}}$ and $\tilde{\sigma}_{\tilde{v}}$ is the same as the subspace spanned by σ_u and σ_v. □

Remark 6.4 With the notation as above,

$$\tilde{\sigma}_{\tilde{u}} \times \tilde{\sigma}_{\tilde{v}} = \det(J)\, \sigma_u \times \sigma_v.$$

Definition 6.5 The unit *normal* to S at Q is defined to be

$$\mathbf{N} = \mathbf{N}_Q = \frac{\sigma_u \times \sigma_v}{\|\sigma_u \times \sigma_v\|},$$

which is independent of the parametrization (up to a sign). Given a smooth parametrization $\sigma : V \to U \subset S \subset \mathbf{R}^3$, the inverse map $\theta : U \to V \subset \mathbf{R}^2$ is called a *chart*, and a collection of charts which cover S, an *atlas*. This idea generalizes to our definition of an abstract surface in Chapter 8.

Example (Sphere $S^2 \subset \mathbf{R}^3$)
- The stereographic projection from the north (respectively south) pole yields two charts on S^2 which form an atlas (see calculations in Section 4.2).
- For the hemisphere $S^2 \cap \{z > 0\}$, we have an obvious chart given by $\theta(x, y, z) = (x, y)$. It is an easy check that there exists an atlas consisting of this and other similar charts.
- Using spherical polar coordinates, there exists an atlas on S^2 consisting of two charts, where the domain of any one is S^2 minus half a great circle. We may for instance take one of these charts to be the inverse of the smooth parametrization σ defined as follows: Let V be the open subset $\{0 < u < \pi,\ 0 < v < 2\pi\}$ of $\mathbf{R}^2$, and define $\sigma : V \to S^2$ by

$$\sigma(u, v) = (\sin u \, \cos v, \sin u \, \sin v, \cos u).$$

It is easy to check (Exercise 6.1) that σ satisfies the defining properties of a smooth parametrization. The other chart will be similar, but will correspond to omitting half a great circle which is disjoint from the first one.

Example (Torus $T \subset \mathbf{R}^3$) There exists a chart on the complement of two appropriate circles on T, where the image is the interior of a square in $\mathbf{R}^2$.

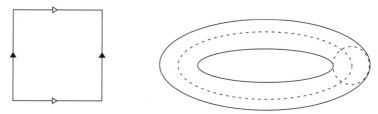

We may use charts which are inverse to the smooth parametrizations given by restricting the map $\sigma : \mathbf{R}^2 \to \mathbf{R}^3$, where

$$\sigma(u, v) = ((2 + \cos u) \cos v, (2 + \cos u) \sin v, \sin u),$$

to different open unit squares in $\mathbf{R}^2$. For appropriate choices of the unit squares, three such charts will suffice to give an atlas.

Definition 6.6   If $S \subset \mathbf{R}^3$ is an embedded surface, the standard inner-product on $\mathbf{R}^3$ restricts to one on the tangent space at any point — this family of inner-products is called the *first fundamental form*.

If $\sigma : V \to U \subset S$ is a parametrization, with $P \in V$, we have induced via $d\sigma_P : \mathbf{R}^2 \xrightarrow{\sim} T_{S,\sigma(P)}$ an inner-product $\langle\ ,\ \rangle_P$ on $\mathbf{R}^2$, varying with P, i.e. a Riemannian metric on V; explicitly,

$$\langle \mathbf{a}, \mathbf{b} \rangle_P = (d\sigma_P(\mathbf{a}), d\sigma_P(\mathbf{b}))_{\mathbf{R}^3}.$$

With respect to the standard basis of $\mathbf{R}^2$, this may be written as

$$E\, du^2 + 2F\, du\, dv + G\, dv^2,$$

where $E = \sigma_u \cdot \sigma_u$, $F = \sigma_u \cdot \sigma_v$, $G = \sigma_v \cdot \sigma_v$; these are clearly smooth functions on V. This Riemannian metric is therefore just the *first fundamental form* expressed in terms of the coordinates (u, v).

Suppose $\tilde{\sigma} = \sigma \circ \phi$ is another parametrization on U, where $\phi : \tilde{V} \to V$ is a diffeomorphism. Given $\mathbf{a}, \mathbf{b} \in \mathbf{R}^2$, we have

$$
\begin{aligned}
&\langle \mathbf{a}, \mathbf{b} \rangle_P^{\sim} && \text{(inner-product for } \tilde{\sigma} \text{ chart)} \\
&= (d\tilde{\sigma}_P(\mathbf{a}), d\tilde{\sigma}_P(\mathbf{b}))_{\mathbf{R}^3} && \text{(inner-product on tangent space)} \\
&= \langle d\phi_P(\mathbf{a}), d\phi_P(\mathbf{b}) \rangle_{\phi(P)} && \text{(inner-product for } \sigma \text{ chart)}
\end{aligned}
$$

since, by the Chain Rule, $d\tilde{\sigma}_P = d\sigma_{\phi(P)} \circ d\phi_P$. Therefore, with respect to the Riemannian metrics on $\tilde{V}$ and V, this says that ϕ is an *isometry*, and so we may apply the relevant results from Chapter 4.

6.2   Lengths and areas

Definition 6.7   Given a smooth curve $\Gamma : [a, b] \to S \subset \mathbf{R}^3$, we define

$$\text{length } \Gamma := \int_a^b \| \Gamma'(t) \|\, dt,$$

$$\text{energy } \Gamma := \int_a^b \| \Gamma'(t) \|^2\, dt.$$

Remark 6.8   The energy should perhaps be called the *action*. Sometimes, in analogy with the formula for kinetic energy, a factor of a half is also included.

Since the image of Γ is a compact set, we may find a dissection of $[a, b]$ which subdivides the curve into finitely many pieces, the image of each of these pieces being contained in a chart — by this, we mean that it is contained in an open set U which admits a chart $\theta : U \to V \subset \mathbf{R}^2$. We can calculate on these charts.

We may therefore reduce to the case when the image of Γ is contained in $U \subset S$, where $\sigma : V \to U$ is a smooth parametrization; the corresponding chart will be denoted by $\theta : U \to V$. We set $\gamma = \theta \circ \Gamma$ to be the corresponding curve in V. Thus $\Gamma = \sigma \circ \gamma$, and

$$
\begin{aligned}
\|\Gamma'(t)\|^2 &= (\Gamma'(t), \Gamma'(t))_{\Gamma(t)} \\
&= (d\sigma_P(\gamma'(t)), d\sigma_P(\gamma'(t)))_{\sigma(P)} \qquad (P = \gamma(t)) \\
&= \langle \gamma'(t), \gamma'(t) \rangle_P = \|\gamma'(t)\|^2,
\end{aligned}
$$

in the norm determined by the Riemannian metric on V.

Writing $\gamma(t) = (\gamma_1(t), \gamma_2(t)) : [a, b] \to \mathbf{R}^2$, this says that

$$
\|\gamma'(t)\| = (E\dot{\gamma}_1^2 + 2F\dot{\gamma}_1\dot{\gamma}_2 + G\dot{\gamma}_2^2)^{1/2}
$$

and

$$
\text{length } \Gamma = \int_a^b (E\dot{\gamma}_1^2 + 2F\dot{\gamma}_1\dot{\gamma}_2 + G\dot{\gamma}_2^2)^{1/2} \, dt.
$$

Since we know how to calculate the length of smooth curves on S, when S is connected we have an associated metric, defined by taking the infimum of lengths of piecewise smooth curves between any two given points (see Section 4.3).

Definition 6.9   Given a smooth parametrization $\sigma : V \to U \subset S$ on an embedded surface $S \subset \mathbf{R}^3$, and an appropriate region $T \subset U$, we define the *area* of T to be the area of $\sigma^{-1}(T)$ with respect to the first fundamental form on V, namely

$$
\int_{\theta(T)} (EG - F^2)^{1/2} \, du \, dv \qquad \text{(when defined)},
$$

where θ denotes the corresponding chart σ^{-1}.

Since

$$
\|\sigma_u \times \sigma_v\|^2 + (\sigma_u \cdot \sigma_v)^2 = \|\sigma_u\|^2 \|\sigma_v\|^2,
$$

for an embedded surface, this may be written as

$$
\int_{\theta(T)} \|\sigma_u \times \sigma_v\| \, du \, dv.
$$

Remark 6.10   Given two parametrizations $\sigma : V \to U$ and $\tilde{\sigma} : \tilde{V} \to U$, we saw above that $\phi = \sigma^{-1} \circ \tilde{\sigma} : \tilde{V} \to V$ is an isometry. Therefore, applying Proposition 4.7, we deduce that the area is defined independent of the parametrization.

This therefore enables us to define areas for more general regions (not necessarily contained in the image of a parametrization). To calculate areas, we often however only need to consider one chart $\theta : U \to V$, since in many cases the subset omitted will not affect the area.

A famous classical result concerning areas is Archimedes theorem; we have already seen an example of this result when we calculated the area of a spherical circle on S^2.

Proposition 6.11 (Archimedes theorem)   *If the sphere S^2 is placed inside a (vertical) circular cylinder of radius one as shown below, then the horizontal radial projection map (with centre the z-axis) from S^2 to the cylinder preserves areas.*

Proof   We take the smooth parametrization $\sigma_1 : V \to U_1 \subset S^2$, where V is the open subset $\{0 < u < \pi,\ 0 < v < 2\pi\}$ of $\mathbf{R}^2$, and

$$\sigma_1(u, v) = (\sin u \, \cos v, \sin u \, \sin v, \cos u).$$

Since the area of a region in S^2 is the same as the area of its intersection with U_1, we can calculate the areas of regions by only using this parametrization.

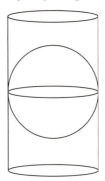

Composing σ_1 with the projection map from the sphere to the cylinder, we obtain a smooth parametrization $\sigma_2 : V \to U_2$, where U_2 is the complement of the line $x = 1, y = 0$ on the cylinder. Explicitly,

$$\sigma_2(u, v) = (\cos v, \sin v, \cos u).$$

We now calculate the first fundamental forms corresponding to these two parametrizations, both of these first fundamental forms being Riemannian metrics on V.

Observing that

$$(\sigma_1)_u = (\cos u \, \cos v, \cos u \, \sin v, -\sin u), \quad \text{and}$$

$$(\sigma_1)_v = (-\sin u \, \sin v, \sin u \, \cos v, 0),$$

we see that $(\sigma_1)_u \cdot (\sigma_1)_u = 1$, $(\sigma_1)_u \cdot (\sigma_1)_v = 0$ and $(\sigma_1)_v \cdot (\sigma_1)_v = \sin^2 u$. The first fundamental form is therefore

$$du^2 + \sin^2 u\, dv^2.$$

Performing the analogous calculations for σ_2, we obtain a first fundamental form

$$\sin^2 u\, du^2 + dv^2.$$

Although these are different metrics, the function $EG - F^2 = \sin^2 u$ in both cases. Thus the area integral we need to perform to calculate the area of a given region $T \subset S^2$ is exactly the same as the area integral we need to perform for the projection of T to the cylinder. □

6.3 Surfaces of revolution

We consider a class of embedded surfaces, on which the calculations are greatly simplified, namely surfaces $S \subset \mathbf{R}^3$ obtained by rotating a plane curve η around a line l. Without loss of generality, we may take l to be the z-axis and assume that the curve η lies in the xz-plane and is given by $\eta : (a, b) \to \mathbf{R}^3$, where

$$\eta(u) = (f(u), 0, g(u))$$

(possibly $a = -\infty$ or $b = \infty$). We assume further that

(i) η is a smoothly immersed curve, that is $\eta'(u) \neq 0$ for all u,
(ii) η is a homeomorphism onto its image (with induced Euclidean metric), and
(iii) $f(u) > 0$ for all u.

The second condition is there to rule out for instance the curves illustrated below. A smoothly immersed curve which satisfies the second condition is called a *smoothly embedded* or *regular* curve.

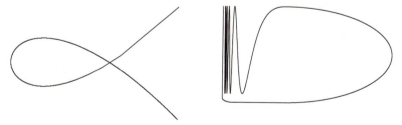

Some books replace the first condition by the assumption that η has unit speed, that is $\|\eta'(u)\| = 1$ for all u. For a smoothly immersed curve, this may always be achieved just by reparametrizing the curve (Lemma 4.3), but in many cases it will be inconvenient to do this. In this chapter, we shall produce formulae which are valid for the more general case, and then comment that these formulae simplify when η has unit speed.

Any point of S is then of the form

$$\sigma(u, v) = (f(u) \cos v, f(u) \sin v, g(u))$$

with $a < u < b, 0 \le v \le 2\pi$. By the assumptions, for any $\alpha \in \mathbf{R}$,

$$\sigma : (a, b) \times (\alpha, \alpha + 2\pi) \to S$$

is a homeomorphism onto its image (note that there exists a continuous choice of the argument on $S^1 \setminus \{e^{i\alpha}\}$). Moreover,

$$\sigma_u = (f' \cos v, f' \sin v, g'),$$
$$\sigma_v = (-f \sin v, f \cos v, 0),$$

and so

$$\sigma_u \times \sigma_v = (-fg' \cos v, -fg' \sin v, ff')$$

and

$$\|\sigma_u \times \sigma_v\|^2 = f^2(f'^2 + g'^2) = f^2 \|\eta'\|^2 \ne 0.$$

Thus (for any α) the map σ is a smooth parametrization, and so S is an embedded surface.

Remark 6.12   We can extend the above definition so as to include embedded surfaces which can be covered by such parametrizations, e.g. the torus and the sphere S^2.

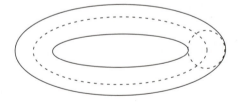
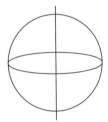

Definition 6.13   Circles obtained by rotating a fixed point of η are called *parallels*, and the curves obtained by rotating the image of η through a fixed angle are called *meridians* (on S^2, these are just the latitudes and longitudes).

The first fundamental form with respect to the parametrization σ is determined by

$$E = \|\sigma_u\|^2 = f'^2 + g'^2,$$
$$F = \sigma_u \cdot \sigma_v = 0,$$
$$G = \|\sigma_v\|^2 = f^2,$$

i.e. takes the form $(f'^2 + g'^2)du^2 + f^2 dv^2$. Under the assumption that η is a unit speed curve, this takes the very simple form $du^2 + f^2 dv^2$. If furthermore f is constant (i.e. S is a circular cylinder), we see that the metric is locally Euclidean.

6.4 Gaussian curvature of embedded surfaces

Before studying the curvature of embedded surfaces, we recall some definitions for embedded curves. Suppose $\eta : [0, l] \to \mathbf{R}^2$ is a smooth curve with unit speed $\|\eta'\| = 1$; at each point $P = \eta(s)$ of the curve, we let $\mathbf{n} = \mathbf{n}(s)$ denote the unit normal to the curve at P, where by convention we assume that the ordered pair of orthonormal vectors $(\eta'(s), \mathbf{n})$ has the same orientation as that of the pair of standard basis vectors (e_1, e_2) of $\mathbf{R}^2$. As $\eta' \cdot \eta' = 1$, we obtain, on differentiating, that $\eta' \cdot \eta'' = 0$; hence $\eta''(s) = \kappa \mathbf{n}$ for some real number κ. Recall that the *curvature* of the curve at the point $\eta(s)$ is then defined to be κ.

For any smooth function $f : [c, d] \to [0, l]$ with $f'(t) > 0$ for all t, we may consider the reparametrized curve $\gamma(t) = \eta(f(t))$. Therefore, with $\dot{\gamma}$ denoting the derivative of γ with respect to t, we have

$$\dot{\gamma}(t) = \frac{df}{dt} \, \eta'(f(t)),$$

and so

$$\|\dot{\gamma}\|^2 = \left(\frac{df}{dt}\right)^2.$$

Now $\eta''(f(t)) = \kappa \, \mathbf{n}$, where κ is the curvature at $\gamma(t)$, and by Taylor's theorem, for small h,

$$\gamma(t + h) - \gamma(t) = \frac{df}{dt} \, \eta'(f(t)) \, h$$
$$+ \frac{1}{2}\left(\left(\frac{d^2 f}{dt^2}\right)\eta'(f(t)) + \left(\frac{df}{dt}\right)^2 \eta''(f(t))\right)h^2 + \cdots.$$

Since $\eta' \cdot \mathbf{n} = 0$, we deduce that

$$(\gamma(t + h) - \gamma(t)) \cdot \mathbf{n} = \frac{1}{2}\kappa \, \|\dot{\gamma}\|^2 h^2 + \cdots.$$

Observe however that

$$\|\gamma(t + h) - \gamma(t)\|^2 = \|\dot{\gamma}\|^2 h^2 + \cdots.$$

Therefore, we have recovered $\frac{1}{2}\kappa$ as the ratio of the quadratic terms of these two expansions, and so κ has been defined independently of the parametrization.

Motivated by this, we consider now the case of embedded *surfaces*. Given $V \subset \mathbf{R}^2$ open and a parametrization $\sigma : V \to U \subset S$, we use Taylor's theorem to expand σ, considered as a vector-valued function, near (u, v).

$$\sigma(u + h, v + k) - \sigma(u, v) = \sigma_u h + \sigma_v k$$
$$+ \frac{1}{2}\left(\sigma_{uu} h^2 + 2\sigma_{uv} hk + \sigma_{vv} k^2\right) + \cdots,$$

noting here the well-known fact that the mixed partial derivatives are symmetric (Theorem 9.34 of [11]).

So the orthogonal deviation of σ from its tangent plane at $P = \sigma(u, v)$ is

$$(\sigma(u + h, v + k) - \sigma(u, v)) \cdot \mathbf{N} = \frac{1}{2}\left(L h^2 + 2M hk + Nk^2\right) + \cdots,$$

where $L = \sigma_{uu} \cdot \mathbf{N}$, $M = \sigma_{uv} \cdot \mathbf{N}$ and $N = \sigma_{vv} \cdot \mathbf{N}$.

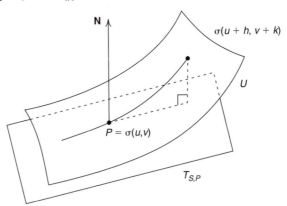

From the definitions of E, F and G, we also have that

$$\|\sigma(u + h, v + k) - \sigma(u, v)\|^2 = Eh^2 + 2Fhk + Gk^2 + \cdots.$$

Definition 6.14   The *second fundamental form* on V is given by the family of bilinear forms

$$L\,du^2 + 2M\,du\,dv + N\,dv^2,$$

where L, M, N are the smooth functions on V defined above. The *Gaussian curvature* K of S at P is defined by

$$K := \frac{LN - M^2}{EG - F^2}.$$

$K > 0$ means that the second fundamental form is positive or negative definite. $K < 0$ means that it is indefinite. $K = 0$ means that it is semi-definite but not definite. We prove below in Corollary 6.18 that the curvature of a surface does not depend on the choice of parametrization.

Example   Given the graph of a smooth function $F(x, y)$ in two variables, the curvature at any given point is given by the value of the *Hessian* $F_{xx}F_{yy} - F_{xy}^2$ at the corresponding point of $\mathbf{R}^2$, scaled by $(1 + F_x^2 + F_y^2)^{-2}$ (Exercise 6.6). A point of $\mathbf{R}^2$ is called a *non-degenerate* point for F if the Hessian does not vanish there. Thus a non-degenerate local maximum or minimum for F gives rise to a point of positive curvature on the graph. On the other hand, a non-degenerate *saddle point* gives rise to a point of negative Gaussian curvature; in one direction the graph curves positively with respect to the normal $\mathbf{N}$ and in the other direction it curves negatively.

A further example of negative curvature is at an "inner point" on an embedded torus. A calculation we perform below gives the curvature at any point on a surface of revolution, which as a special case yields the curvature on the torus.

Example   Consider a circular cylinder in $\mathbf{R}^3$. Here, there is a locally Euclidean metric, induced from the locally Euclidean metric on the 'unfolded surface'.

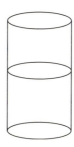

Let us take a parametrization given by

$$\sigma(u, v) = (\cos v, \sin v, u)$$

with $-\infty < u < \infty$, $\alpha \leq v \leq \alpha + 2\pi$. Therefore,

$$\sigma_u = (0, 0, 1),$$
$$\sigma_v = (-\sin v, \cos v, 0),$$

and so the first fundamental form is $du^2 + dv^2$, as previously calculated. The second fundamental form is readily calculated to be dv^2. Thus, at any point of the cylinder, the second fundamental form is non-zero, but the curvature is zero.

There is a useful alternative definition for the functions L, M and N in the second fundamental form.

Lemma 6.15   *With the notation as above, the unit normal can be regarded as a smooth vector-valued function $\mathbf{N}(u, v)$ of the variables u, v. We then have that $-L = \sigma_u \cdot \mathbf{N}_u$, $-M = \sigma_u \cdot \mathbf{N}_v = \sigma_v \cdot \mathbf{N}_u$, and $-N = \sigma_v \cdot \mathbf{N}_v$.*

Proof   We note that $\sigma_u \cdot \mathbf{N} = 0$ and $\sigma_v \cdot \mathbf{N} = 0$. Differentiating these equations with respect to u and v, we obtain the claimed identities.   □

Proposition 6.16   *In the above notation, if the second fundamental form is identically zero on V, and V is connected, then $\sigma(V)$ is an open subset of a plane in $\mathbf{R}^3$.*

Proof   If the second fundamental form is zero, then the previous lemma implies that $\mathbf{N}_u$ and $\mathbf{N}_v$ are orthogonal to both σ_u and σ_v. Since $\mathbf{N}$ is a unit vector, we have $\mathbf{N}{\cdot}\mathbf{N} = 1$; differentiating with respect to u and v shows that $\mathbf{N}_u$ and $\mathbf{N}_v$ are also orthogonal to $\mathbf{N}$, and hence that $\mathbf{N}_u = 0$ and $\mathbf{N}_v = 0$. A componentwise use of the Mean Value theorem implies that $\mathbf{N}$ is locally constant, and then the connectedness of V implies that $\mathbf{N}$ is a constant vector. Considering σ as a vector-valued function of u, v, we deduce that $\sigma(u, v) \cdot \mathbf{N}$ is a constant, since we get zero when we differentiate with respect to u or v. Hence U is contained in some plane, given by $\mathbf{x} \cdot \mathbf{N} = $ constant.   □

There is another useful characterization of the curvature; it is this result which yields the fact that the curvature is independent of any choice of parametrization.

Proposition 6.17   *If $\mathbf{N}$ denotes the unit normal of the surface patch σ, i.e.*

$$\mathbf{N} = \frac{\sigma_u \times \sigma_v}{\|\sigma_u \times \sigma_v\|},$$

then at a given point,

$$\mathbf{N}_u = a\sigma_u + b\sigma_v,$$
$$\mathbf{N}_v = c\sigma_u + d\sigma_v,$$

where

$$-\begin{pmatrix} L & M \\ M & N \end{pmatrix} = \begin{pmatrix} a & b \\ c & d \end{pmatrix}\begin{pmatrix} E & F \\ F & G \end{pmatrix}.$$

In particular, $K = ad - bc$.

Proof   Since $\mathbf{N} \cdot \mathbf{N} = 1$, we have $\mathbf{N}_u \cdot \mathbf{N} = 0$ and $\mathbf{N}_v \cdot \mathbf{N} = 0$. So $\mathbf{N}_u$ and $\mathbf{N}_v$ are in the tangent space at P, and may be written in the form

$$\mathbf{N}_u = a\sigma_u + b\sigma_v,$$
$$\mathbf{N}_v = c\sigma_u + d\sigma_v,$$ (†)

for some $\begin{pmatrix} a & b \\ c & d \end{pmatrix}$. But $\mathbf{N}_u \cdot \sigma_u = -L$, $\mathbf{N}_u \cdot \sigma_v = \mathbf{N}_v \cdot \sigma_u = -M$ and $\mathbf{N}_v \cdot \sigma_v = -N$. Taking the dot product of (†) with σ_u and σ_v gives

$$-L = aE + bF, \qquad -M = aF + bG,$$
$$-M = cE + dF, \qquad -N = cF + dG,$$

i.e.

$$-\begin{pmatrix} L & M \\ M & N \end{pmatrix} = \begin{pmatrix} a & b \\ c & d \end{pmatrix}\begin{pmatrix} E & F \\ F & G \end{pmatrix}.$$

Taking determinants, the final claim follows.   □

Corollary 6.18   *K is independent of the parametrization.*

Proof By Proposition 6.17, $\mathbf{N}_u \times \mathbf{N}_v = K\sigma_u \times \sigma_v$. Suppose we reparametrize U by means of a diffeomorphism $\phi : \tilde{V} \to V$.

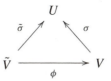

We have seen in Remark 6.4 that $\tilde{\sigma}_{\tilde{u}} \times \tilde{\sigma}_{\tilde{v}} = \det(J)\,\sigma_u \times \sigma_v$, where $J = J(\phi)$ is the Jacobian matrix, and so $\tilde{\mathbf{N}} = \pm\mathbf{N}$, depending on the sign of $\det J$. In particular, $\tilde{\mathbf{N}}_{\tilde{u}} \times \tilde{\mathbf{N}}_{\tilde{v}} = \mathbf{N}_{\tilde{u}} \times \mathbf{N}_{\tilde{v}}$.

By the Chain Rule,

$$\mathbf{N}_{\tilde{u}} = \frac{\partial u}{\partial \tilde{u}}\mathbf{N}_u + \frac{\partial v}{\partial \tilde{u}}\mathbf{N}_v,$$

$$\mathbf{N}_{\tilde{v}} = \frac{\partial u}{\partial \tilde{v}}\mathbf{N}_u + \frac{\partial v}{\partial \tilde{v}}\mathbf{N}_v,$$

and so

$$\mathbf{N}_{\tilde{u}} \times \mathbf{N}_{\tilde{v}} = \det(J)\,\mathbf{N}_u \times \mathbf{N}_v.$$

Therefore

$$\det(J)K\,\sigma_u \times \sigma_v = \det(J)\,\mathbf{N}_u \times \mathbf{N}_v$$
$$= \mathbf{N}_{\tilde{u}} \times \mathbf{N}_{\tilde{v}}$$
$$= \tilde{K}\tilde{\sigma}_{\tilde{u}} \times \tilde{\sigma}_{\tilde{v}}$$
$$= \tilde{K}\det(J)\,\sigma_u \times \sigma_v,$$

and thus $\tilde{K} = K$ as claimed. □

A geometrically appealing result on curvature is the following.

Proposition 6.19 *If S is an embedded surface in $\mathbf{R}^3$ which is closed and bounded (i.e. compact), then the Gaussian curvature must be strictly positive at some point of S.*

Proof Since S is compact, there exists a point P of S whose Euclidean distance from the origin in $\mathbf{R}^3$ is a maximum. We choose a smooth parametrization $\sigma : V \to U \ni P$, for some open subset V of $\mathbf{R}^2$, say with $\sigma(u_0, v_0) = P$. We use Taylor's theorem to expand σ, considered as a vector-valued function, near (u_0, v_0).

$$\sigma(u_0 + h, v_0 + k) = \sigma(u_0, v_0) + \sigma_u h + \sigma_v k$$
$$+ \frac{1}{2}(\sigma_{uu} h^2 + 2\sigma_{uv} hk + \sigma_{vv} k^2) + O(3).$$

Thus

$$\|\sigma(u_0 + h, v_0 + k)\|^2 = \|\sigma(u_0, v_0)\|^2 + 2\sigma(u_0, v_0) \cdot (\sigma_u h + \sigma_v k) + O(2).$$

Since the left-hand side of this last equation is less than or equal to the constant term on the right-hand side for all small values of h and k, we must have that $\sigma(u_0, v_0) \cdot \sigma_u = 0$ and $\sigma(u_0, v_0) \cdot \sigma_v = 0$. If $\mathbf{N}$ denotes the normal to S at P, this implies that $\sigma(u_0, v_0) = \lambda\mathbf{N}$ for some non-zero $\lambda \in \mathbf{R}$ — clearly, we cannot have P being at the origin, and hence λ is non-zero.

We now expand further, obtaining

$$\|\sigma(u_0 + h, v_0 + k)\|^2 - \|\sigma(u_0, v_0)\|^2 = E\,h^2 + 2F\,hk + G\,k^2$$
$$+ \lambda(L\,h^2 + 2M\,hk + N\,k^2) + O(3).$$

Therefore, the quadratic form associated to the matrix

$$\begin{pmatrix} \lambda L + E & \lambda M + F \\ \lambda M + F & \lambda N + G \end{pmatrix}$$

(evaluated at (u_0, v_0)) is negative semi-definite. As the matrix $\left(\begin{smallmatrix} E & F \\ F & G \end{smallmatrix}\right)$ is positive definite, we deduce that the matrix $\lambda\left(\begin{smallmatrix} L & M \\ M & N \end{smallmatrix}\right)$ is negative definite, and hence that the matrix $\left(\begin{smallmatrix} L & M \\ M & N \end{smallmatrix}\right)$ is definite (either positive or negative, according to the sign of λ). This in turn is just the condition that $LN - M^2$ is strictly positive, and hence that the curvature is strictly positive at P. $\qquad\qquad\square$

Thus, if for instance we consider the locally Euclidean torus T, introduced as a metric space in Chapter 3, we may deduce that it cannot be realized as the metric space underlying an embedded surface in $\mathbf{R}^3$. If an embedded surface S has a first fundamental form which is locally Euclidean, then the curvature vanishes — here, we implicitly use Corollary 8.2 below. If however T is realized as an embedded surface, then (as it is compact) the previous result says that the curvature would have to be strictly positive at some point, yielding a contradiction.

Finally, let us return to the case of surfaces of revolution; here it is easy to calculate curvature.

Proposition 6.20   *With the notation for surfaces of revolution as before, the Gaussian curvature is given by the formula*

$$K = \frac{(f'g'' - f''g')g'}{f(f'^2 + g'^2)^2}.$$

In the case when the curve η has unit speed, this takes the form $K = -f''/f$.

Proof   Recall that locally we have a smooth parametrization of the form

$$\sigma : (a, b) \times (\alpha, \alpha + 2\pi) \to U \subset S,$$

where
$$\sigma(u,v) = (f(u)\cos v, f(u)\sin v, g(u)).$$

Moreover, as previously calculated,
$$\sigma_u = (f'\cos v, f'\sin v, g'),$$
$$\sigma_v = (-f\sin v, f\cos v, 0),$$

and the first fundamental form is $(f'^2 + g'^2)du^2 + f^2 dv^2$.
    We also calculated that

$$\sigma_u \times \sigma_v = (-fg'\cos v, -fg'\sin v, ff'),$$

and that
$$\|\sigma_u \times \sigma_v\|^2 = f^2(f'^2 + g'^2).$$

The unit normal vector is

$$\mathbf{N} = (-g'\cos v, -g'\sin v, f')/(f'^2 + g'^2)^{1/2}.$$

However

$$\sigma_{uu} = (f''\cos v, f''\sin v, g''),$$
$$\sigma_{uv} = (-f'\sin v, f'\cos v, 0),$$
$$\sigma_{vv} = (-f\cos v, -f\sin v, 0),$$

from which it follows that

$$L = (f'g'' - f''g')/(f'^2 + g'^2)^{1/2}, \quad M = 0, \quad \text{and} \quad N = fg'/(f'^2 + g'^2)^{1/2}.$$

The curvature therefore is

$$K = \frac{LN - M^2}{EG - F^2} = \frac{(f'g'' - f''g')g'}{f(f'^2 + g'^2)^2}.$$

When $f'^2 + g'^2 = 1$, this takes the form $(f'g'g'' - f''g'^2)/f$. Differentiating $f'^2 + g'^2 = 1$, we obtain $g'g'' = -f'f''$; substituting this into the previous expression, and using again the identity $f'^2 + g'^2 = 1$, we obtain the simplification claimed. $\square$

Example   As examples of the last result, let us consider the sphere and the embedded torus. The unit sphere is the surface of revolution corresponding to the curve $\eta : (0, \pi) \to \mathbf{R}^3$ given by $\eta(u) = (\sin u, 0, \cos u)$; we remark that the $f'^2 + g'^2 = 1$ condition is satisfied. Since $f(u) = \sin u$, we have $K = -f''/f = 1$ at all points.

    The embedded torus is the surface of revolution corresponding to the unit circle with centre $(2, 0, 0)$, which may be parametrized as $\eta : (\alpha, \alpha + 2\pi) \to \mathbf{R}^3$ (for

α a real number), where $\eta(u) = (2 + \cos u, 0, \sin u)$; again we observe that the $f'^2 + g'^2 = 1$ condition is satisfied. Here, $f(u) = 2 + \cos u$, and so the curvature is $K = \cos u / (2 + \cos u)$. In particular, it is positive for $-\pi/2 < u < \pi/2$ (the 'outer' points of the torus), is negative for $\pi/2 < u < 3\pi/2$ (the 'inner' points), and is zero on the two circles given by $x^2 + y^2 = 1$, $z = \pm 1$.

We shall see in Chapter 8 that the curvature depends only on the first fundamental form (i.e. the metric), and does not otherwise depend on the embedding. Given this fact, we can identify the isometry group of the embedded torus, since an isometry must then preserve curvature. Given the calculation above, it follows easily that the group of direct isometries of an embedded torus is just S^1, corresponding to the rotations about the z-axis. The group of all isometries contains S^1 as an index two subgroup, and may be thought of as a continuous version of the dihedral group. Unlike the case of the locally Euclidean torus, the action of the isometry group is clearly not transitive.

Exercises

6.1 Let V be the open subset $\{0 < u < \pi, \ 0 < v < 2\pi\}$ in $\mathbf{R}^2$, and let $\sigma : V \to S^2$ be given by

$$\sigma(u, v) = (\sin u \ \cos v, \sin u \ \sin v, \cos u).$$

Prove that σ defines a smooth parametrization of a certain open subset of S^2. [You may assume that $\cos^{-1}$ is continuous on $(-1, 1)$, and that $\tan^{-1}, \cot^{-1}$ are continuous on $(-\infty, \infty)$.]

6.2 Let $\gamma : [0, 1] \to \mathbf{R}^2$ be a regular simple closed plane curve, given by $\gamma(u) = (\gamma_1(u), \gamma_2(u))$. Let S be the image of $V = [0, 1] \times \mathbf{R}$ under the map $(u, v) \mapsto (\gamma_1(u), \gamma_2(u), v)$. Show that S is an embedded surface, and that, with respect to suitable parametrizations, the first fundamental form corresponds to the Euclidean metric on $\mathbf{R}^2$.

6.3 With S denoting the embedded surface from the previous question, show that S is isometric to a circular cylinder of radius $\text{length}(\gamma)/2\pi$.

6.4 Let T denote the embedded torus in $\mathbf{R}^3$ obtained by rotating around the z-axis the circle $(x - 2)^2 + z^2 = 1$ in the xz-plane. Using the formal definition of area in terms of a parametrization, calculate the surface area of T.

6.5 Sketch the embedded surface in $\mathbf{R}^3$ given by the equation

$$(x^2 + y^2)(z^4 + 1) = 1,$$

and show that it has bounded area.

6.6 Let $S \subset \mathbf{R}^3$ denote the graph of a smooth function F (defined on some open subset of $\mathbf{R}^2$), given therefore by the equation $z = F(x, y)$. Show that S is a smooth embedded surface, and that its curvature at a point $(x, y, z) \in S$ is the value taken at (x, y) by

$$(F_{xx}F_{yy} - F_{xy}^2)/(1 + F_x^2 + F_y^2)^2.$$

6.7 Show that the curvature of the embedded hyperboloid of two sheets, with equation $x^2 + y^2 = z^2 - 1$ in $\mathbf{R}^3$, is everywhere positive. [Compare this result with the calculation in Section 5.7.]

6.8 Sketch the surface $S \subset \mathbf{R}^3$ given by $z = \exp(-(x^2 + y^2)/2)$, and find a formula for its Gaussian curvature at a general point. Show that the curvature is strictly positive at a point $(x, y, z) \in S$ if and only if $x^2 + y^2 < 1$.

6.9 Let $S \subset \mathbf{R}^3$ be the ellipsoid $x^2/a^2 + y^2/b^2 + z^2/c^2 = 1$. If $V \subset \mathbf{R}^2$ denotes the region $u^2/a^2 + v^2/b^2 < 1$, show that the map

$$\sigma(u, v) = \left(u, \, v, \, c\,(1 - u^2/a^2 - v^2/b^2)^{1/2} \right)$$

determines a smooth parametrization of a certain open subset of S. Prove that the Gaussian curvatures at the points $(a, 0, 0), (0, b, 0), (0, 0, c)$ are all equal if and only if $a = b = c$, i.e. S is a sphere.

6.10 If $S \subset \mathbf{R}^3$ is a surface of revolution with curvature everywhere zero, show that it is an open subset of either a plane, a circular cylinder or a circular cone. In each case, find local coordinates with respect to which the metric is Euclidean.

6.11 Let $f(u) = e^u$, $g(u) = (1 - e^{2u})^{1/2} - \cosh^{-1}(e^{-u})$, where $u < 0$, and S be the surface of revolution corresponding to the curve $\eta : (-\infty, 0) \to \mathbf{R}^3$ given by $\eta(u) = (f(u), 0, g(u))$. Show that S has constant Gaussian curvature -1; S is called the *pseudosphere*. By considering coordinates v and $w = e^{-u}$ on S, show that the pseudosphere is isometric to the open subset of the upper half-plane model of the hyperbolic plane given by $\mathrm{Im}(z) > 1$.

[By a theorem of Hilbert, the hyperbolic plane cannot itself be realized as an embedded surface.]

6.12 Suppose that $S \subset \mathbf{R}^3$ is a surface of revolution with constant curvature one, which may be compactified to a smooth closed embedded surface by the addition of precisely two further points. Show that S is a unit sphere, minus two antipodal points.

6.13 Let $f(x, y, z)$ be a smooth real-valued function on $\mathbf{R}^3$, and let $S \subset \mathbf{R}^3$ denote its zero locus, given by $f = 0$. Suppose P is a point of S at which $\partial f/\partial z(P) \neq 0$; show that the map $\mathbf{R}^3 \to \mathbf{R}^3$ given by

$$(x, y, z) \mapsto (x, y, f(x, y, z))$$

is a local diffeomorphism at P. Hence show that there exists a smooth parametrization of some open neighbourhood of P in S.

Suppose now one knows that the differential df_P is non-zero for all $P \in S$; prove that S is an embedded surface in the sense of Definition 6.1. For $P \in S$, identify the tangent space at P as a certain codimension one subspace of $\mathbf{R}^3$. [Such a surface S is called an *unparametrized* smooth embedded surface in $\mathbf{R}^3$.]

7 Geodesics

In the specific geometries we studied in earlier chapters (Euclidean, spherical, hyperbolic, ...), the concept of *lines* proved central, as did their property of being (locally) length minimizing with respect to the relevant metric. In this chapter, we generalize these ideas and obtain the concept of *geodesic curves* on a general surface. It turns out to be simpler to approach this via the energy of a curve rather than its length, but we shall see in Section 7.3 that the two approaches are closely related. The property that a smooth curve is *geodesic* is in fact a local one, and this observation always enables us to reduce down to the case of an open subset $V \subset \mathbf{R}^2$, equipped with a Riemannian metric. We therefore study this case first.

7.1 Variations of smooth curves

Suppose V is an open subset of $\mathbf{R}^2$ with coordinates (u, v), and equipped with a Riemannian metric

$$E \, du^2 + 2F \, du \, dv + G \, dv^2.$$

Given a smooth curve $\gamma : [a, b] \to V$, we can write it in terms of coordinates $\gamma(t) = (u(t), v(t))$ and define its *energy* by the formula

$$\text{energy } \gamma = \int_a^b (E(u, v)\dot{u}^2 + 2F(u, v)\dot{u}\dot{v} + G(u, v)\dot{v}^2) \, dt,$$

where as usual $\dot{u}$ denotes du/dt and $\dot{v}$ denotes dv/dt. This is consistent with Definition 6.7, and the remark following that definition is relevant here also. To simplify the notation (as well as dealing with a more general problem), we write this integral as

$$\int_a^b I(t, u, v, \dot{u}, \dot{v}) \, dt,$$

where in our specific case I only depends on t through its dependence on $u, \dot{u}, v, \dot{v}$.

Readers who are already familiar with the theory of Calculus of Variations may wish to pass briefly over the remainder of this section. For the convenience of other readers however, we sketch the relevant theory below.

Definition 7.1   Given a smooth curve $\gamma : [a, b] \rightarrow V$, a *variation* of γ is given by a smooth map $h : [a, b] \times (-\varepsilon, \varepsilon) \rightarrow V$, the left-hand side being a subset of $\mathbf{R}^2$, such that $h(t, 0) = \gamma(t)$ for all $t \in [a, b]$. It is called a *proper variation* if the end-points are fixed under the variation, namely $h(a, \tau) = \gamma(a)$ and $h(b, \tau) = \gamma(b)$ for all $\tau \in (-\varepsilon, \varepsilon)$. For each $\tau \in (-\varepsilon, \varepsilon)$, we have a smooth curve $\gamma_\tau : [a, b] \rightarrow V$ given by $\gamma_\tau(t) = h(t, \tau)$.

Remark   In certain circumstances, we might wish to extend this definition by allowing h to be only continuous in the variable τ, and where (for fixed τ) it is piecewise smooth in the variable t, but here we shall restrict to the smooth case.

We wish to know how the integral

$$\int_a^b I(t, u, v, \dot{u}, \dot{v}) \, dt,$$

changes under small variations and in particular under small proper variations.

With the notation as above, suppose we consider a nearby curve γ_τ in the variation. We can write

$$\gamma_\tau(t) = (u(t) + \delta u(t), v(t) + \delta v(t)),$$

for appropriate smooth functions δu and δv of t (depending also on τ). Assuming that I is a smooth function in each of its variables (which it patently is in the case of the energy functional), the integral at τ may be written, up to first order terms, as

$$\int_a^b \frac{\partial I}{\partial u} \delta u \, dt + \int_a^b \frac{\partial I}{\partial \dot{u}} \delta \dot{u} \, dt + \int_a^b \frac{\partial I}{\partial v} \delta v \, dt + \int_a^b \frac{\partial I}{\partial \dot{v}} \delta \dot{v} \, dt$$

Since

$$\frac{\partial I}{\partial \dot{u}} \delta \dot{u} = \frac{d}{dt} \left(\frac{\partial I}{\partial \dot{u}} \delta u \right) - \delta u \frac{d}{dt} \left(\frac{\partial I}{\partial \dot{u}} \right),$$

with a similar identity for the variable v, we deduce that the integral may be written (up to first order terms) as

$$\int_a^b \left(\frac{\partial I}{\partial u} - \frac{d}{dt} \left(\frac{\partial I}{\partial \dot{u}} \right) \right) \delta u \, dt + \int_a^b \left(\frac{\partial I}{\partial v} - \frac{d}{dt} \left(\frac{\partial I}{\partial \dot{v}} \right) \right) \delta v \, dt$$

$$+ \left[\frac{\partial I}{\partial \dot{u}} \delta u \right]_a^b + \left[\frac{\partial I}{\partial \dot{v}} \delta v \right]_a^b. \tag{7.1}$$

Writing $h(t, \tau) = (u(t, \tau), v(t, \tau))$ and taking limits as $\tau \rightarrow 0$, we find that the derivative (with respect to τ) of the integral, at $\tau = 0$, is given by

$$\int_a^b \left(\frac{\partial I}{\partial u} - \frac{d}{dt} \left(\frac{\partial I}{\partial \dot{u}} \right) \right) \frac{\partial u}{\partial \tau} \, dt + \int_a^b \left(\frac{\partial I}{\partial v} - \frac{d}{dt} \left(\frac{\partial I}{\partial \dot{v}} \right) \right) \frac{\partial v}{\partial t} \, dt$$

$$+ \left[\frac{\partial I}{\partial \dot{u}} \frac{\partial u}{\partial \tau} \right]_a^b + \left[\frac{\partial I}{\partial \dot{v}} \frac{\partial v}{\partial \tau} \right]_a^b. \tag{7.2}$$

If this derivative is zero, we say that γ represents a *stationary point* for the integral with respect to the given variation.

A particular case occurs when the variation is proper, and so for all τ we have $\delta u(a) = \delta u(b) = 0$, $\delta v(a) = \delta v(b) = 0$; hence

$$\frac{\partial u}{\partial \tau}(a) = \frac{\partial u}{\partial \tau}(b) = 0, \qquad \frac{\partial v}{\partial \tau}(a) = \frac{\partial v}{\partial \tau}(b) = 0.$$

The third and fourth terms in Equation (7.2) do not then occur.

Proposition 7.2 (Calculus of variations)   *The integral*

$$\int_a^b I(t, u, v, \dot{u}, \dot{v}) \, dt$$

is stationary at γ for all possible proper variations if and only if the Euler–Lagrange *equations*

$$\frac{d}{dt}\left(\frac{\partial I}{\partial \dot{u}}\right) = \frac{\partial I}{\partial u}, \qquad \frac{d}{dt}\left(\frac{\partial I}{\partial \dot{v}}\right) = \frac{\partial I}{\partial v}$$

are satisfied for all $t \in (a, b)$.

Proof   From Equation (7.2), it is clear that the integral is stationary at γ for all proper variations if the Euler–Lagrange equations hold; this is simply the statement that the two integrals in (7.2) vanish.

For the converse, we observe that for any smooth curve $\eta = (\eta_1, \eta_2) : [a, b] \to \mathbf{R}^2$ with $\eta(a) = (0, 0) = \eta(b)$, we can consider the proper variation of γ given by $\gamma_\tau = \gamma + \tau\eta$, for τ sufficiently small. Thus, in Equation (7.2), the functions $\frac{\partial u}{\partial \tau}$ and $\frac{\partial v}{\partial \tau}$ may be chosen to be arbitrary smooth functions of $t \in [a, b]$ vanishing at a and b. In particular we may take $\frac{\partial u}{\partial \tau}$ to be the smooth function

$$\frac{\partial I}{\partial u} - \frac{d}{dt}\left(\frac{\partial I}{\partial \dot{u}}\right)$$

multiplied by an appropriate smooth bump function to force vanishing at the endpoints (such a bump function will take value 1 on $[a + \varepsilon, b - \varepsilon]$ for some $\varepsilon > 0$ sufficiently small, but will vanish at a and b). From this, the reader may check easily that the first Euler–Lagrange equation holds; the second equation follows similarly.
$\square$

The Calculus of Variations and the Euler–Lagrange equations are central to much of both physics and geometry. Here is not the place to go into details concerning this statement — here, we shall be interested in the application of the theory to curves representing stationary points of length or energy, namely the geodesic curves as defined below. Before doing this, we shall however digress to give another application, applying the above theory to find stationary points of the area functional for surfaces of revolution.

Example (Minimal surfaces of revolution)   We consider surfaces of revolution S in $\mathbf{R}^3$ given by an equation of the form $x^2 + y^2 = f(z)^2$, for $a < z < b$, where $f : [a,b] \to \mathbf{R}$ is a strictly positive smooth function. We let $\eta : (a,b) \to \mathbf{R}^3$ be the smooth embedded curve in the xz-plane defined by $\eta(t) = (f(t),0,t)$ — the fact that it is a smooth embedded curve being an easy exercise. The surface S is then the surface of revolution determined by this curve, and its closure has boundary given by the two circles C_1 and C_2, with equations $z = a$, $x^2 + y^2 = f(a)^2$, and $z = b$, $x^2 + y^2 = f(b)^2$, respectively. For fixed values of $f(a)$ and $f(b)$ (and hence fixed boundary circles C_1 and C_2), we seek to find the function f (and hence the embedded curve η), representing a stationary point for the area of the embedded surface S. We call such a surface a *minimal surface of revolution*, and it will turn out to be unique; one can show that it may be represented physically by a soap film stretched between the fixed circles C_1 and C_2. The fact that we have taken the curve η in an apparently special form, namely the graph of the function f, does not significantly affect the final conclusion (Exercise 7.9).

Using our calculation from the previous chapter of the first fundamental form for S, one checks easily that, for a given smooth function $f : [a,b] \to \mathbf{R}$, the area of S is given by the formula

$$\int_a^b I(f,f')\,dt \quad \text{where} \quad I(f,f') = f(f'^2 + 1)^{1/2}.$$

If, for given boundary conditions, we wish to find the function f for which the area is stationary, we need to solve the single Euler–Lagrange equation (writing u for the function f)

$$\frac{d}{dt}\left(\frac{\partial I}{\partial \dot{u}}\right) = \frac{\partial I}{\partial u}.$$

Since however I does not depend directly on t, we have the equation

$$\frac{dI}{dt} = \dot{u}\frac{\partial I}{\partial u} + \ddot{u}\frac{\partial I}{\partial \dot{u}},$$

from which it follows that the Euler–Lagrange equation may be rewritten as

$$\frac{d}{dt}\left(I - \dot{u}\frac{\partial I}{\partial \dot{u}}\right) = 0.$$

This is just the statement that the bracket, which in the case under consideration is readily seen to be $f/(f'^2 + 1)^{1/2}$, is a constant.

If we set this (positive) constant to be $1/c$, then the function f satisfies the differential equation $f' = ((cf)^2 - 1)^{1/2}$, which may be solved to give f in the form

$$f(t) = \frac{1}{c}\cosh(ct + k).$$

The constants c and k just need to be chosen so that $f(a)$ and $f(b)$ are the given fixed values.

The reader may recognise $y = \frac{1}{c}\cosh(ct + k)$ as the equation of a *catenary*, the curve formed by a chain of uniform density, with fixed end-points, hanging under its own weight. For this reason, the above minimal surface is often called a *catenoid*.

Let us now return to the case of main interest for us, namely the energy functional (for a given Riemannian metric $E\,du^2 + 2F\,du\,dv + G\,dv^2$ on some open subset V of $\mathbf{R}^2$). We saw that the integrand for the energy could be written as

$$I(u, v, \dot{u}, \dot{v}) = E(u, v)\dot{u}^2 + 2F(u, v)\dot{u}\dot{v} + G(u, v)\dot{v}^2.$$

For the case of the energy functional therefore, we have

$$\frac{\partial I}{\partial \dot{u}} = 2(E\dot{u} + F\dot{v}), \qquad \frac{\partial I}{\partial u} = E_u\dot{u}^2 + 2F_u\dot{u}\dot{v} + G_u\dot{v}^2,$$

and analogous identities for $\frac{\partial I}{\partial \dot{v}}$ and $\frac{\partial I}{\partial v}$. Suppose then that $\gamma : [a, b] \to V$ is a smooth curve, and we write $\gamma(t) = (\gamma_1(t), \gamma_2(t))$. By Proposition 7.2, saying that γ represents a stationary point for the energy with respect to all proper variations is equivalent to the Euler–Lagrange equations

$$\frac{d}{dt}(E\dot{\gamma_1} + F\dot{\gamma_2}) = \frac{1}{2}(E_u\dot{\gamma_1}^2 + 2F_u\dot{\gamma_1}\dot{\gamma_2} + G_u\dot{\gamma_2}^2)$$
$$\frac{d}{dt}(F\dot{\gamma_1} + G\dot{\gamma_2}) = \frac{1}{2}(E_v\dot{\gamma_1}^2 + 2F_v\dot{\gamma_1}\dot{\gamma_2} + G_v\dot{\gamma_2}^2) \tag{7.3}$$

being satisfied for all $t \in (a, b)$.

Definition 7.3   A smooth curve $\gamma : [a, b] \to V$ is called *geodesic* if the above Euler–Lagrange equations hold. These ordinary differential equations are therefore known as the *geodesic equations*. With this definition, it is clear that the property of being a geodesic is a purely local condition on the curve.

For the Euclidean plane $\mathbf{R}^2$, we have $E = G = 1$ and $F = 0$, and so the geodesic equations reduce to $\ddot{\gamma_1} = 0 = \ddot{\gamma_2}$. Thus the geodesics are the curves γ with $\ddot{\gamma} = 0$, that is just the lines in $\mathbf{R}^2$, parametrized with constant speed. As a slightly less trivial example, we prove directly from the geodesic equations that the geodesics in the hyperbolic plane are just the hyperbolic lines, parametrized with constant speed. This latter result could in fact be deduced without further calculation using the length-minimizing property of hyperbolic lines, proved in Proposition 5.8, and the general results contained in the next three sections.

Lemma 7.4   *The geodesics in the hyperbolic plane are precisely the hyperbolic lines parametrized with constant speed.*

Proof   We shall take the upper half-plane model H of the hyperbolic plane, equipped with the hyperbolic metric $(dx^2 + dy^2)/y^2$. Thus $E = G = 1/y^2$ and $F = 0$. We show that if $\gamma : [0, 1] \to H$ is a geodesic curve joining two points, then it is a hyperbolic line segment parametrized with constant speed, and the converse is similar. Applying an appropriate isometry, we may assume that the two points are on the imaginary axis. We set $\gamma(t) = u(t) + i\,v(t)$; the geodesic equations then take the form

$$\frac{d}{dt}\left(2\dot{u}/v^2\right) = 0, \qquad \frac{d}{dt}\left(2\dot{v}/v^2\right) = -2(\dot{u}^2 + \dot{v}^2)/v^3.$$

The first equation implies that $\dot{u} = cv^2$ for some constant c; thus, if $c \neq 0$, we have that $\dot{u}$ always has the same sign. Since, by assumption $u(0) = u(1) = 0$, we deduce that $c = 0$ and $\dot{u} = 0$.

The second equation then reduces to $\ddot{v}/v^2 = \dot{v}^2/v^3$, or $v\ddot{v} = \dot{v}^2$, since $v(t) > 0$ for all t. Since

$$\frac{d}{dt}\left(\frac{\dot{v}}{v}\right) = (v\ddot{v} - \dot{v}^2)/v^2 = 0,$$

we deduce that $\dot{v}/v$ is constant. Since $\|\dot{\gamma}\|^2 = \dot{v}^2/v^2$, this is just the statement that $\|\dot{\gamma}\|$ is constant. $\qquad\square$

7.2   Geodesics on embedded surfaces

Suppose $S \subset \mathbf{R}^3$ is an embedded surface and $\sigma : V \to U \subset S$ a parametrization, with $\theta = \sigma^{-1}$ the corresponding chart. If $\Gamma : [a,b] \to U$ is a smooth curve on S, then $\gamma = \theta \circ \Gamma$ is a smooth curve on V.

We saw in the previous chapter that

$$\|\Gamma'(t)\|^2 = \|\gamma'(t)\|^2,$$

where the right-hand side was calculated using the corresponding Riemannian metric on V; the energy may if we wish therefore be calculated on the chart. Because it is the same as the energy of Γ, it does not depend on the choice of parametrization $\sigma : V \to U$. In the next chapter, we shall introduce more general *abstract surfaces* defined in terms of charts, and there the energy of a curve must be defined by means of the charts. The compatibility conditions we shall introduce between different charts will however again ensure that the definition is independent of any choice of chart.

We say that the curve Γ *locally* represents a stationary point of the energy under proper variations if, for any $t_0 \in (a,b)$, there exists $\varepsilon > 0$ sufficiently small such that $\Gamma|_{[t_0-\varepsilon,\,t_0+\varepsilon]}$ represents a stationary point for the energy of curves joining $\Gamma(t_0 - \varepsilon)$ to $\Gamma(t_0 + \varepsilon)$. Now Γ locally represents a stationary point of the energy under proper variations if and only if γ locally represents a stationary point of the energy under proper variations, the energy here being calculated via the Riemannian metric on V determined by the first fundamental form. This latter condition holds if and only if the geodesic equations hold locally. We may therefore define Γ to be *geodesic* if γ is geodesic, and the above interpretation in terms of being locally a stationary point for the energy ensures that this definition does not depend on the choice of chart.

For the case of an embedded surface, the above discussion enables us to define when an arbitrary smooth curve $\Gamma : [a,b] \to S$ is geodesic. This is a purely local definition: for a given t, we choose a local parametrization $\sigma : V \to U \ni \Gamma(t)$ and demand that Γ, and hence also γ, are geodesics locally.

Corollary 7.5 *If a curve Γ on an embedded surface S minimizes the energy functional for curves joining $P = \Gamma(a)$ and $Q = \Gamma(b)$, then it is a geodesic in the above sense.*

Proof For any $a < a_1 < b_1 < b$, the curve $\Gamma_1 = \Gamma|_{[a_1,b_1]}$ minimizes the energy functional for smooth curves joining $\Gamma(a_1)$ to $\Gamma(b_1)$. If there were a smooth curve joining $\Gamma(a_1)$ to $\Gamma(b_1)$ for which the energy were smaller, then it would be easy to construct a piecewise smooth curve from $\Gamma(a)$ to $\Gamma(b)$ which had strictly smaller energy than Γ, and this could be smoothed at a_1, b_1 to achieve a smooth curve with the same property.

If a_1, b_1 are chosen such that Im Γ_1 is contained in a chart U, then Γ_1 is certainly a stationary point for the energy under proper variations, and hence is a geodesic. By varying a_1, b_1, the corollary follows. $\quad\square$

Remark The proof of Corollary 7.5 shows that if Γ *locally* minimizes the energy, then Γ is a geodesic. The *converse* is also true — geodesics *locally* minimize energy (see Corollary 7.18).

There is another interpretation of the geodesic Equations (7.3) for the case of an embedded surface $S \subset \mathbf{R}^3$; this interpretation corresponds to the statement in the Euclidean plane that the geodesics are curves with zero acceleration, namely lines parametrized with constant speed. In the general case of an arbitrary Riemannian metric, one has a concept of the *covariant derivative* of a vector field along a curve, which we shall not define here (in the embedded case, it is just the projection of the ordinary derivative to the tangent plane). The corresponding condition for a curve to be geodesic is that the covariant derivative of the tangent field to the curve is identically zero.

Proposition 7.6 *For a smooth curve Γ on an embedded surface S, the geodesic equations are equivalent to the statement that $\frac{d^2\Gamma}{dt^2}$ is always normal to S.*

Proof The result being local, we may reduce to the case when $\Gamma : [a, b] \to U \subset S$ with $\sigma : V \to U$ a parametrization; then $\Gamma = \sigma \circ \gamma$ with $\gamma(t) = \gamma_1(t)e_1 + \gamma_2(t)e_2$ and hence, by the Chain Rule,

$$\dot{\Gamma}(t) = (d\sigma)_{\gamma(t)}\dot{\gamma}(t)$$
$$= (d\sigma)_{\gamma(t)}(\dot{\gamma}_1(t)e_1 + \dot{\gamma}_2(t)e_2)$$
$$= \dot{\gamma}_1(t)\,\sigma_u + \dot{\gamma}_2(t)\,\sigma_v.$$

So $\frac{d^2\Gamma}{dt^2}$ is perpendicular to the subspace $\langle\sigma_u, \sigma_v\rangle \subset \mathbf{R}^3$ spanned by σ_u, σ_v if and only if

$$\sigma_u \cdot \frac{d}{dt}(\dot{\gamma}_1\,\sigma_u + \dot{\gamma}_2\,\sigma_v) = 0,$$
$$\sigma_v \cdot \frac{d}{dt}(\dot{\gamma}_1\,\sigma_u + \dot{\gamma}_2\,\sigma_v) = 0,$$

for all t. The first of these equations may be written as

$$0 = \frac{d}{dt}((\dot{\gamma}_1\sigma_u + \dot{\gamma}_2\sigma_v) \cdot \sigma_u) - (\dot{\gamma}_1\sigma_u + \dot{\gamma}_2\sigma_v) \cdot \frac{d\sigma_u}{dt}$$

$$= \frac{d}{dt}(E\dot{\gamma}_1 + F\dot{\gamma}_2) - (\dot{\gamma}_1\sigma_u + \dot{\gamma}_2\sigma_v) \cdot (\dot{\gamma}_1\sigma_{uu} + \dot{\gamma}_2\sigma_{uv}).$$

Since

$$E_u = (\sigma_u \cdot \sigma_u)_u = 2\sigma_u \cdot \sigma_{uu},$$

$$F_u = (\sigma_u \cdot \sigma_v)_u = \sigma_u \cdot \sigma_{uv} + \sigma_v \cdot \sigma_{uu},$$

$$G_u = (\sigma_v \cdot \sigma_v)_u = 2\sigma_v \cdot \sigma_{uv},$$

we observe that the first equation above is equivalent to the first of the geodesic Equations (7.3). In a similar way, the second equation is equivalent to the second geodesic equation. $\square$

Remark 7.7   Therefore, if Γ is a geodesic on an embedded surface S, then $\frac{d}{dt}(\dot{\Gamma} \cdot \dot{\Gamma}) = 2\dot{\Gamma} \cdot \ddot{\Gamma} = 0$ and so $\|\frac{d\Gamma}{dt}\|^2$ is *constant*. More generally, the geodesic Equations (7.3) imply directly that $\|\dot{\gamma}\|^2$ is constant — for a non-illuminating proof of this, see [9], p. 181, Exercise 8.12 (answer on p. 310). There is a better proof that the speed is constant in the general case, which is formally the same as that given here in the embedded case, but using the covariant derivative (mentioned above) instead of d/dt.

7.3   Length and energy

Firstly, we recall the Cauchy–Schwarz inequality for integrals. If f, g are continuous real-valued functions on $[a, b]$, then

$$\left(\int_a^b fg\right)^2 \le \int_a^b f^2 \int_a^b g^2$$

with equality if and only if $f = 0$ or $g = \lambda f$ for some $\lambda \in \mathbf{R}$.

Given a Riemannian metric on an open subset $V \subset \mathbf{R}^2$ and a smooth curve $\gamma : [a, b] \to V$, we may apply this with $f = 1, g = \|\dot{\gamma}\|$ to give

$$(\text{length } \gamma)^2 \le (b - a)\,\text{energy } \gamma$$

with equality if and only if $\|\dot{\gamma}\|$ is constant.

Lemma 7.8   *Let $V \subset \mathbf{R}^2$ be an open subset equipped with a Riemannian metric. We consider smooth curves $\gamma : [0, 1] \to V$ joining $P = \gamma(0)$ to $Q = \gamma(1)$. A curve γ_0 minimizes the energy if and only if it minimizes the length and has constant speed.*

Proof   For such curves γ, we have $(\text{length } \gamma)^2 \le \text{energy } \gamma$, with equality if and only if $\|\dot{\gamma}\|$ is constant. For a given length l, the minimum energy is l^2 (achieved when the speed is constant). So γ_0 minimizes the energy if and only if it minimizes the length and has constant speed. □

Corollary 7.9 *If a smooth curve Γ on an embedded surface $S \subset \mathbf{R}^3$ has constant speed and locally minimizes the length, then it is a geodesic.* □

The difference between a smoothly embedded curve locally minimizing the length and locally minimizing the energy is essentially just a matter of the parametrization, as by Lemma 4.3 any smooth curve with nowhere vanishing derivative may be reparametrized so as to have constant speed. We recall that a curve $\gamma(t)$ is parametrized with speed one if and only if the parameter t is (modulo an additive constant) just the arc-length.

Example At this stage, it is perhaps helpful to give an example of a geodesic which does not minimize the energy. By the preceding discussion, it follows that on the sphere S^2, segments of great circles parametrized with constant speed are geodesics. Let us consider two non-antipodal points on S^2, and the segment γ of the great circle between them of larger length. The reader is invited to convince herself that there exists a proper variation of γ under which the length and energy decrease, and one under which the length and energy increase. From these, it is possible also to construct a proper variation for which γ represents a stationary point which is neither a maximum nor a minimum.

7.4 Existence of geodesics

The existence locally of geodesics through a given point, that is, solutions to the geodesic equations, follows from standard facts concerning the existence of solutions for systems of ordinary differential equations.

Proposition 7.10 *Given an open subset $V \subset \mathbf{R}^2$, equipped with a Riemannian metric, $P = (u_0, v_0) \in V$ and $(p_0, q_0) \in \mathbf{R}^2$, there exists precisely one geodesic $\gamma : (-\varepsilon, \varepsilon) \to V$ with $\gamma(0) = P$ and $\gamma'(0) = (p_0, q_0)$.*

Corollary 7.11 *Through each point P on an embedded surface S and for each direction at P, there is a unique (germ of a) geodesic.* □

Proof of Proposition 7.10 Using coordinates (u, v) on $V \subset \mathbf{R}^2$, the second order non-linear differential Equations (7.3) say that $E\ddot{u} + F\ddot{v}$ and $F\ddot{u} + G\ddot{v}$ are functions of $u, v, \dot{u}, \dot{v}$. Using the fact that

$$\begin{pmatrix} E & F \\ F & G \end{pmatrix}$$

is invertible, the geodesic equations may therefore be written in the form

$$\ddot{u} = f(u, v, \dot{u}, \dot{v}),$$
$$\ddot{v} = g(u, v, \dot{u}, \dot{v}),$$

or equivalently, as a first order system,

$$\dot{u} = p$$
$$\dot{v} = q$$
$$\dot{p} = f(u, v, p, q)$$
$$\dot{q} = g(u, v, p, q).$$

Standard theory of ordinary differential equations implies that there exists a unique solution on $(-\varepsilon, \varepsilon)$ for some $\varepsilon > 0$, with initial conditions (u_0, v_0, p_0, q_0).   □

Remark   A slightly stronger form of the local existence theorem for systems of ordinary differential equations ensures that, since the functions f and g are smooth, the local solutions depend smoothly on the initial parameters u_0, v_0, p_0, q_0. We return to this point below, in Theorem 7.13.

Example   We have seen that segments of great circles on the sphere (with constant speed parametrizations) are geodesics. Proposition 7.10 implies that these are the *only* geodesics, since through each $P \in S^2$, there exists a unique great circle in any given direction. Similarly, we deduce that the segments of hyperbolic lines are the *only* geodesics on the hyperbolic plane, hence completing an alternative proof of Lemma 7.4.

Example   Consider a circular cylinder in $\mathbf{R}^3$. Here, the geodesics correspond to any straight line segments on the 'unfolded surface' with the Euclidean metric. To see this algebraically, take the smooth parametrization

$$\sigma(u, v) = (\cos v, \sin v, u)$$

with $-\infty < u < \infty, \alpha < v < \alpha + 2\pi$. The first fundamental form is just $du^2 + dv^2$, and so is locally Euclidean. Therefore $\gamma(t) = (u(t), v(t))$ is a geodesic if and only if $\ddot{u} = 0, \ddot{v} = 0$, that is, u and v are linear in t as claimed.

We now revisit the commonly occuring class of embedded surfaces introduced in Section 6.3, on which the calculations were greatly simplified, namely surfaces of revolution $S \subset \mathbf{R}^3$. Recall that these were obtained by rotating a *smoothly embedded* plane curve η around a line l. Without loss of generality, we took l to be the z-axis and assumed the curve η to lie in the xz-plane, given by $\eta : (a, b) \to \mathbf{R}^3$, where

$$\eta(u) = (f(u), 0, g(u)),$$

with $f(u) > 0$ for all u.

Now we assume also that η is parametrized so as to have unit speed, so that $\|\eta'(u)\| = 1$ for all u — recall that, by Lemma 4.3, this may always be achieved after a smooth reparametrization. Any point of S is of the form

$$\sigma(u, v) = (f(u) \cos v, f(u) \sin v, g(u))$$

with $a < u < b, 0 \leq v < 2\pi$. We saw that, for any $\alpha \in \mathbf{R}$,

$$\sigma : (a, b) \times (\alpha, \alpha + 2\pi) \to S$$

is a smooth parametrization, and that the first fundamental form with respect to this parametrization takes the form $du^2 + f^2 dv^2$.

The geodesic equations for a curve $\gamma(t) = (u(t), v(t))$ then read

$$\ddot{u} = f(u)\frac{df}{du}\dot{v}^2, \qquad \frac{d}{dt}\left(f(u)^2\dot{v}\right) = 0 \qquad (7.4)$$

(and equivalently for the curve $\Gamma = \sigma \circ \gamma$).

If γ is a geodesic, we know from Remark 7.7 that it has constant speed, and, after rescaling the parameter, we may assume that $\|\dot{\gamma}\| = 1$, i.e.

$$\boxed{\dot{u}^2 + f(u)^2\dot{v}^2 = 1}$$

Proposition 7.12   *Consider the surface of revolution determined as above by a unit speed curve η, with resulting parametrization*

$$\sigma(u, v) = (f(u)\cos v, f(u)\sin v, g(u)).$$

(i)   *Every unit speed meridian is a geodesic,*

(ii)   *A (unit speed) parallel $u = u_0$ is a geodesic if and only if $\frac{df}{du}(u_0) = 0$.*

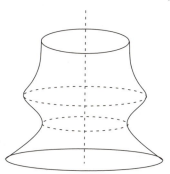

Proof

(i)   As v is constant on a meridian, the second equation of (7.4) is satisfied. By the unit speed condition, $\dot{u}$ is constant, so the first equation is also satisfied.

(ii)   Given that $u = u_0$ on a unit speed parallel, the unit speed equation for γ reads $f(u)^2\dot{v}^2 = 1$, implying that $\dot{v} = \pm 1/f(u_0)$, a non-zero constant. Therefore, the second equation of (7.4) clearly holds. The first equation holds if and only if $\frac{df}{du}(u_0) = 0$.   □

Remark   We remark that for u_0 as in (ii), we know that $|\frac{dg}{du}(u_0)| = 1$, and so locally f may be regarded as a function of z. The condition given is then just saying that the point in question is a stationary point of the graph, as illustrated above.

Example   On the embedded torus $T \subset \mathbf{R}^3$, Proposition 7.12 yields some canonical geodesics. With the torus embedded as illustrated in Remark 6.12, all the vertical circles on T are geodesics, but the only horizontal circles that are geodesics are those of minimum and maximum radius. The general geodesic on T is not in fact even a closed curve. This is a non-obvious fact, dating back to the work of Darboux — it is however easy to prove the corresponding result for the torus equipped with its locally Euclidean metric, for which the geodesics correspond locally to line segments in the chart defined by an open fundamental square in $\mathbf{R}^2$, and the question of whether the geodesic is closed or not then reduces down to whether the gradient of the line is rational or irrational.

7.5   Geodesic polars and Gauss's lemma

Let us consider first the familiar case of polar coordinates on $\mathbf{R}^2 = \mathbf{C}$. Any non-zero point of $\mathbf{R}^2$ has unique polar coordinates (r, θ), with $r > 0$ and $0 \le \theta < 2\pi$. Moreover, given any angle θ_0, there is a smooth parametrization of $U = \mathbf{C} \setminus \mathbf{R}_{\ge 0} e^{i\theta_0}$

$$\sigma : (0, \infty) \times (\theta_0, \theta_0 + 2\pi) \to U$$

given by $\sigma(r, \theta) = (r \cos \theta, r \sin \theta)$, with inverse (chart) given locally by

$$(x, y) \to ((x^2 + y^2)^{1/2}, \tan^{-1}(y/x)).$$

The second coordinate here needs to be interpreted appropriately, writing the function as $\cot^{-1}(x/y)$ when $x = 0$, and choosing the correct value locally for $\tan^{-1}(y/x)$, respectively $\cot^{-1}(x/y)$. Note that the radial rays given by setting θ constant are just the geodesics starting at the origin, and that the radial coordinate r at a point is the distance along the relevant geodesic.

For an arbitrary surface, we can construct, at least locally, a similar coordinate system. In fact, it is sufficient to do this for an open subset $V \subset \mathbf{R}^2$ equipped with a Riemannian metric. In the case of an embedded surface, we can find a chart $\psi : U \to V$ with $P \in U$, and V an open subset of $\mathbf{R}^2$ equipped with the Riemannian metric given by the first fundamental form. In the case of an abstract surface, as introduced in the next chapter, such a chart to an open subset V of $\mathbf{R}^2$ equipped with a Riemannian metric exists from the definition. The geodesics through P on U will then correspond under ψ to geodesics through $\psi(P)$ in V, and the construction given below will yield immediately the required coordinate system on the corresponding open subset of U.

We suppose therefore that $V \subset \mathbf{R}^2$ is an open subset, equipped with a Riemannian metric, and assume that $P \in V$. It follows from Proposition 7.10 that for any angle θ, there is a unique (germ of a) geodesic

$$\gamma_\theta : (-\varepsilon, \varepsilon) \to V$$

through P with $\|\dot\gamma_\theta(0)\| = 1$ and whose tangent at P has polar angle θ. Recall from Remark 7.7 that geodesics have constant speed, and so $\|\dot\gamma_\theta(t)\| = 1$ for all $-\varepsilon < t < \varepsilon$.

We have already remarked that the geodesic γ_θ will depend smoothly on the initial parameter θ — rather more is however true.

Theorem 7.13

(i) *For fixed $P \in V$, we may choose $\varepsilon > 0$ (independent of θ) for which the geodesics $\gamma_\theta : (-\varepsilon, \varepsilon) \to V$ are defined on $(-\varepsilon, \varepsilon)$ for all θ. Moreover, if we vary $P \in V$, we may take ε to be a continuous (in fact smooth) function of P.*

(ii) *Let B_ε denote the ε-ball centred at the origin in $\mathbf{R}^2$, and define a map $\sigma : B_\varepsilon \to V$ by $\sigma(r \cos\theta, r \sin\theta) := \gamma_\theta(r)$ (with $\sigma(\mathbf{0}) = P$). The map σ is smooth, and, for ε sufficiently small, it is a diffeomorphism from B_ε onto an open neighbourhood of W of P in V.*

Proof These results follow from an analysis of the differential equations involved, which the reader is invited to take on trust. Once we know that σ is smooth, we may proceed as follows: Given $\mathbf{v}(\theta) = (\cos\theta, \sin\theta) \in \mathbf{R}^2$, we can consider the line segment $\eta(t) = t\,\mathbf{v}(\theta)$, for $-\varepsilon < t < \varepsilon$. By definition, we have $\gamma_\theta(t) = \sigma(\eta(t))$, and so by the Chain Rule

$$(d\sigma)_\mathbf{0}(\mathbf{v}(\theta)) = \dot\gamma_\theta(0) \neq 0.$$

Since this holds for all θ, we have $(d\sigma)_\mathbf{0}$ is an isomorphism; the final claim then follows from the Inverse Function theorem. A full reference for these results is [12], Chapter 9. □

Definition 7.14 An open neighbourhood of P of the form $W = \sigma(B_\varepsilon)$, as defined in Theorem 7.13(ii), is called a *normal neighbourhood*. We observe that, for any $Q \in W \setminus \{P\}$, there exists a unique geodesic in W from P to Q.

Choosing $\varepsilon > 0$ small enough therefore, we can define a smooth map $g : (-\varepsilon, \varepsilon) \times \mathbf{R} \to V$ by

$$g(r, \theta) := \gamma_\theta(r) = \sigma(r\,\mathbf{v}(\theta)),$$

where $\mathbf{v}(\theta) = (\cos\theta, \sin\theta)$. For any angle θ_0, this restricts to a diffeomorphism of $(0, \varepsilon) \times (\theta_0, \theta_0 + 2\pi)$ onto an open subset of V. Note that the image of $(0, \varepsilon) \times (\theta_0, \theta_0 + 2\pi)$ under g is not a neighbourhood of P. The image of $(0, \varepsilon) \times [\theta_0, \theta_0 + 2\pi)$ is the punctured neighbourhood $W \setminus \{P\}$ of P, where $W = \sigma(B_\varepsilon)$ is the normal neighbourhood of P. The coordinates (r, θ) constructed here are called *geodesic polar coordinates* around P. Often one uses ρ instead of r, to indicate that this coordinate comes from the length of the geodesic ray. A point of $W \setminus \{P\}$ will have well-defined geodesic polar coordinates (r, θ) with $r > 0$ and $0 \le \theta < 2\pi$.

With respect to geodesic polar coordinates, the geodesic rays in V starting from P correspond to taking θ to be constant, and the 'geodesic circles' centred at P (of radius $< \varepsilon$) correspond to taking r to be constant.

Theorem 7.15 (Gauss's lemma) *The curves given by taking r to be a positive constant $< \varepsilon$ intersect all the geodesic rays through P at right angles.*

Proof For fixed $r < \varepsilon$, consider the smooth curve α in W given by

$$\alpha(\tau) = (\alpha_1(\tau), \alpha_2(\tau)) := \sigma(r \cos\tau, r \sin\tau).$$

For a given value of τ, we have a geodesic ray $\sigma_\tau(t) := \sigma(tr \cos \tau, tr \sin \tau)$, i.e. in our previous notation $\gamma_\tau(tr)$, where $0 \le t \le 1$. Setting

$$h(t, \tau) := \sigma(tr \cos \tau, tr \sin \tau)$$

for $0 \le t \le 1$ and $\tau \in \mathbf{R}$, we obtain a variation of the geodesic $\gamma = \sigma_0$ by geodesics σ_τ which have fixed initial point P but variable end-point, namely $\alpha(\tau)$. We let $Q = \alpha(0) = \gamma(1)$, the point where the geodesic ray γ meets the curve α. Each σ_τ has length r and constant speed r, so by Lemma 7.8 the energy of σ_τ is also constant as τ varies.

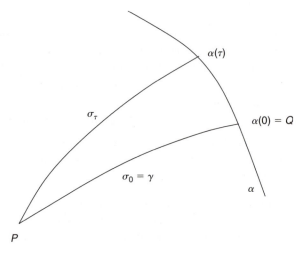

Using the formula (7.2) for the derivation of the energy at $\tau = 0$, with $a = 0$, $b = 1$ and $I = E(u, v)\dot{u}^2 + 2F(u, v)\dot{u}\dot{v} + G(u, v)\dot{v}^2$, we deduce that

$$0 = \left(E(Q)\frac{d\gamma_1}{dt}(1) + F(Q)\frac{d\gamma_2}{dt}(1) \right) \frac{d\alpha_1}{d\tau}(0)$$
$$+ \left(F(Q)\frac{d\gamma_1}{dt}(1) + G(Q)\frac{d\gamma_2}{dt}(1) \right) \frac{d\alpha_2}{d\tau}(0),$$

the two integrals in Equation (7.2) not appearing since γ is a geodesic. This equation is just the statement

$$\left\langle \frac{d\gamma}{dt}(1), \frac{d\alpha}{d\tau}(0) \right\rangle_Q = 0,$$

which we needed to prove. $\square$

Given now the diffeomorphism $\sigma : B_\varepsilon \to W$ as above, we defined a smooth map $g : (-\varepsilon, \varepsilon) \times \mathbf{R} \to W$ by $g(t, \theta) = \sigma(t\,\mathbf{v}(\theta))$, where $\mathbf{v}(\theta) = (\cos \theta, \sin \theta) \in \mathbf{R}^2$. This restricts to a smooth map $h : (0, \varepsilon) \times \mathbf{R} \to W$, given by the same formula, and we have an induced Riemannian metric on $(0, \varepsilon) \times \mathbf{R}$, essentially just the Riemannian metric from W expressed in terms of the geodesic polar coordinates (r, θ). We denote this metric by $E\,dr^2 + 2F\,dr\,d\theta + G\,d\theta^2$. Since the length of a geodesic ray is given by the coordinate r, it follows that $E = 1$. Gauss's lemma says that $F = 0$. The Riemannian metric is therefore expressed, with respect to these geodesic polar coordinates, as

$$dr^2 + G(r,\theta)\,d\theta^2$$

The function $G(r,\theta)$ is a smooth (positive) function on $(0,\varepsilon) \times \mathbf{R}$, periodic in θ with period 2π. Since $h = (h_1, h_2)$ is locally an isometry (by construction), we deduce that $G(r,\theta) = \|dh(e_2)\|^2 = \|\partial h/\partial\theta\|^2$, where e_1, e_2 is the standard basis for $\mathbf{R}^2$, and the norm is determined by the Riemannian metric at $h(r,\theta)$. This however shows that we may extend G to a smooth (positive) function on $(-\varepsilon,\varepsilon) \times \mathbf{R}$, by setting

$$G(t,\theta) = \left\| \frac{\partial g}{\partial \theta}(t,\theta) \right\|^2_{g(t,\theta)}.$$

Lemma 7.16   *With $G(t,\theta)$ the function on $(-\varepsilon,\varepsilon) \times \mathbf{R}$ defined above, we have $G(t,\theta) = t^2 q(t,\theta)$ for some positive smooth function q on $(-\varepsilon,\varepsilon) \times \mathbf{R}$ with $q(0,\theta) = 1$ for all θ.*

Proof   Recalling that $g(t,\theta) = \sigma(t\,\mathbf{v}(\theta)) = \gamma_\theta(t)$, we have

$$\frac{\partial g}{\partial \theta}(t,\theta) = (d\sigma)_{t\mathbf{v}}(t\,\mathbf{v}'(\theta)),$$

where $\mathbf{v}'(\theta) = (-\sin\theta, \cos\theta) = \mathbf{v}(\theta + \pi/2)$. Therefore

$$G(t,\theta) = \|\partial g/\partial\theta\|^2 = t^2 q(t,\theta),$$

where $q(t,\theta) = \|(d\sigma)_{t\mathbf{v}}(\mathbf{v}'(\theta))\|^2$ is a smooth function, strictly positive for $t \neq 0$. To find the value of q at $t = 0$, we set $\phi = \theta + \pi/2$ and observe that

$$q(0,\theta) = \|(d\sigma)_0(\mathbf{v}(\phi))\|^2_P = \|\dot\gamma_\phi(0)\|^2 = 1. \qquad \square$$

Remark 7.17   Considering $G(r,\theta)$ for $r > 0$, this result determines the initial asymptotics of G as $r \to 0$. It follows immediately for instance that $G(r,\theta) \to 0$ and $G_r/r \to 2$ as $r \to 0$. Moreover, since $t\,q(t,\theta)^{1/2}$ is also a smooth function on $(-\varepsilon,\varepsilon) \times \mathbf{R}$, it follows that $(\sqrt{G})_r \to 1$ as $r \to 0$.

We now observe that the above formula for the metric, in terms of geodesic polar coordinates, enables us to tie up a loose end from earlier in this chapter. The proof given shows that the result is equally true for geodesics on embedded (or, more generally, abstract) surfaces.

Corollary 7.18   *With the notation as above, the radial geodesic curve from P to $\sigma(r_0, \theta_0) = Q$ of length $r_0 < \varepsilon$ represents an absolute minimum for the length of curves joining P to Q. More generally, geodesics always locally minimize length and energy.*

Proof The above formula for the metric in terms of geodesic polar coordinates enables us to estimate the length of curves from below. Suppose that $\gamma(t)$ is a smooth curve joining P to Q, which is contained in $W = \sigma(B_\varepsilon)$; then

$$\text{length } \gamma = \int (\dot{r}^2 + G\dot{\theta}^2)^{1/2} \, dt \geq \int |\dot{r}| \, dt \geq r_0,$$

with equality if and only if $\dot{\theta} = 0$ and $r(t)$ is monotonic (cf. proof of Proposition 5.8). On the other hand, any smooth curve $\gamma(t)$ joining P to Q, which does not remain in W, must have length at least ε (cf. proof of Lemma 4.4). Hence the geodesic ray segment from P to Q given by $\theta = \theta_0$ represents an absolute minimum for the lengths of curves joining P to Q.

In general, we reduce locally to the case of the geodesic lying in a normal neighbourhood of its initial point, and hence representing an absolute minimum for both the length and energy (using Lemma 7.8). □

Thus in Gauss's lemma, the curves given by taking r to be a positive constant $< \varepsilon$ consist of the points whose geodesic distance from P is r, and so are indeed *geodesic circles* centred at P. We note in passing that geodesic circles (however parametrized) are usually not geodesics (Exercise 7.7).

To illustrate the above ideas, we return to our three classical geometries, and calculate their metrics in terms of geodesic polar coordinates, which we shall now denote as (ρ, θ).

Example

(i)   For the Euclidean plane, $\mathbf{R}^2$, the geodesic polar coordinates at the origin coincide with the standard polar coordinates (r, θ), and with respect to these coordinates the metric is $d\rho^2 + \rho^2 \, d\theta^2$: thus $G(\rho, \theta) = \rho^2$.

(ii) For the sphere S^2, with respect to geodesic polar coordinates at a point, which we may for instance take to be the north pole, the metric takes the form $d\rho^2 + \sin^2 \rho \, d\theta^2$. For a formal proof of this, we can use the geodesic chart $\sigma(\rho, \theta) = (\sin \rho \cos \theta, \sin \rho \sin \theta, \cos \rho)$. Therefore $G(\rho, \theta) = \sin^2 \rho$.

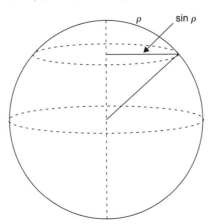

(iii) On the disc model of the hyperbolic plane, the metric is

$$\left(\frac{2}{1-r^2}\right)^2 (dr^2 + r^2\, d\theta^2),$$

with respect to the standard polar coordinates. Consider now geodesic polar coordinates (ρ,θ) at the origin, where $\rho = 2\tanh^{-1} r$. Therefore, $d\rho^2 = \left(\frac{2}{1-r^2}\right)^2 dr^2$. But $r = \tanh\frac{1}{2}\rho$, and so $\frac{4r^2}{(1-r^2)^2} = \sinh\rho$; thus the metric (in geodesic polar coordinates) is $d\rho^2 + \sinh^2\rho\, d\theta^2$. Therefore $G(\rho,\theta) = \sinh^2\rho$.

This yields yet another model of the (punctured) hyperbolic plane, namely $\mathbf{R}^2$ with Riemannian metric $dr^2 + \sinh^2 r\, d\theta^2$, defined in terms of the standard polar coordinates; by changing coordinates, we see that this also defines a metric at the origin.

Finally, we prove an easy but important local result concerning the shape of geodesics in a normal neighbourhood of a point. Consider a normal neighbourhood $W = \sigma(B_\delta)$ of a point P; to fix ideas, we may as before reduce to the case when $P \in V$, an open subset of $\mathbf{R}^2$ equipped with a Riemannian metric, although the argument is equally applicable to the case of an embedded or abstract surface. With respect to geodesic polar coordinates on $W \setminus \{P\}$, the metric may be expressed as $dr^2 + G(r,\theta)d\theta^2$. We recall that $G_r/r \to 2$ as $r \to 0$.

Lemma 7.19 *Suppose that $Q_1, Q_2 \in W$, and that $\gamma(t)$ is a geodesic segment in W joining the two points. If $G_r > 0$ at all points of this geodesic segment, then the maximum distance from P to a point on the geodesic is attained at either Q_1 or Q_2. If $G_r < 0$ at all points of this geodesic, then the minimum distance from P to a point on the geodesic is attained at either Q_1 or Q_2.*

Proof Since $G_r/r \to 2$ as $r \to 0$, the geodesic segment will not contain P.

We may assume therefore that the geodesic segment is contained in $W \setminus \{P\}$; we shall be interested in the local form of γ, and so we can write γ in terms of geodesic polar coordinates $\gamma(t) = (r(t),\theta(t))$; thus $||\dot{\gamma}||^2 = \dot{r}^2 + G\dot{\theta}^2$. The first of the geodesic equations is now of the form

$$\ddot{r} = \frac{1}{2}G_r\dot{\theta}^2.$$

So if $t = t_0$ represents a stationary point for r, we have $\dot{r} = 0$ and $\dot{\theta}^2 > 0$. In the case when $G_r > 0$, we cannot have a local maximum of $r(t)$ at $t = t_0$; when $G_r < 0$, we cannot have a local minimum. The lemma therefore follows. $\square$

If one flies from New York to Moscow, along the (shorter) great circle route, one's distance from the north pole is maximum at take-off, and minimum at some point during the journey. If, on the other hand, one flies from Rio de Janeiro to Sydney, Australia along the (shorter) great circle route, one's distance from the north pole is minimum at take-off, and maximum at some point during the journey. This follows from the above calculation, together with a rudimentary knowledge of geography, since if we take geodesic polar coordinates (ρ, θ) at the north pole, then $G(r, \theta) = \sin^2 \rho$; thus $G_\rho = 2 \cos \rho \sin \rho = \sin 2\rho$ is positive in the northern hemisphere ($\rho < \pi/2$), and negative in the southern hemisphere. In this case however, a simple geometric argument (which we saw in the proof of Proposition 2.16) also suffices to prove these facts.

Exercises

7.1 Show that any line on an embedded surface $S \subset \mathbf{R}^3$ must be a geodesic. Hence, find infinitely many geodesics on the hyperboloid of one sheet, with equation $x^2 + y^2 = z^2 + 1$, in addition to those obtained via Proposition 7.12.

7.2 Suppose we have a Riemannian metric on an open disc D of radius $\delta > 0$ centred on the origin in $\mathbf{R}^2$, possibly with D being all of $\mathbf{R}^2$, given in standard polar coordinates by

$$(dr^2 + r^2 \, d\theta^2)/h(r)^2,$$

where $h(r) > 0$ for all $0 \le r < \delta$. Write down the geodesic equations for this metric. Show that any radial curve, parametrized so as to have unit speed, is a geodesic.

7.3 With the Riemannian metric as in the previous question, show directly, without using the geodesic equations, that the length minimizing curves through the origin are just the line segments which contain the origin.

7.4 For $a > 0$, let $S \subset \mathbf{R}^3$ be the circular half-cone defined by $z^2 = a(x^2 + y^2), z > 0$, considered as an embedded surface. Show that the metric on S is locally Euclidean. When $a = 3$, give an explicit description of the geodesics and show that no geodesic intersects itself. For $a > 3$, prove that there are geodesics (of infinite length) which intersect themselves.

7.5 Let $S \subset \mathbf{R}^3$ be an embedded surface and $H \subset \mathbf{R}^3$ a plane which is normal to S (i.e. contains the unit normal vector) at each point of the intersection $C = S \cap H$. Suppose γ is a constant speed curve on S whose image is contained in C; deduce from Proposition 7.6 that γ is a geodesic on S.

7.6 Using Proposition 7.6, provide an alternative proof of Proposition 7.12.

7.7 Let V denote an open subset of $\mathbf{R}^2$ equipped with a Riemannian metric, and suppose that we have geodesic polar coordinates (r, θ) at a point $P \in V$, for $r < \varepsilon$. For a fixed $r_0 < \varepsilon$, suppose that the function $G(r, \theta)$ defined above has $G_r(r_0, \theta) = 0$ for only finitely many $0 \le \theta < 2\pi$. Show that the geodesic circle centred on P with radius r_0 is not a geodesic. Give an example of a geodesic circle which is a geodesic.

7.8 Consider the Riemannian metric on the unit disc in $\mathbf{R}^2$ defined by

$$\frac{1}{1 - r^2}(dr^2 + r^2 \, d\theta^2),$$

with respect to the standard polar coordinates. Express the metric in terms of the corresponding geodesic polar coordinates (centred on the origin).

7.9 Let $\eta(t) = (f(t), 0, g(t)) : [a, b] \to \mathbf{R}^3$ be a smooth embedded curve in the xz-plane, with $f(t) > 0$ for all t, and let S denote the surface of revolution defined by η, with boundary consisting of the two circles C_1 and C_2, with equations $z = g(a)$,

$x^2 + y^2 = f(a)^2$, respectively $z = g(b)$, $x^2 + y^2 = f(b)^2$. Suppose now that η represents a stationary point for the area of S; if $g(a) \neq g(b)$, prove that S is a catenoid, as described in Section 7.1. What happens if $g(a) = g(b)$?

7.10   Suppose V is an open subset of $\mathbf{R}^2$, equipped with a Riemannian metric, whose associated metric space is *complete*. Show that any geodesic $\gamma : (-\varepsilon, \varepsilon) \to V$ may be extended to a *complete* geodesic, that is a geodesic $\gamma : \mathbf{R} \to V$. Show that the same fact holds for complete embedded surfaces. [The converse is also true in both cases, and follows from the *Hopf–Rinow theorem*.]

8 Abstract surfaces and Gauss–Bonnet

8.1 Gauss's Theorema Egregium

Suppose S is an embedded surface; in the previous chapter we proved the existence locally of *geodesic polar coordinates*, and saw that, with respect to these coordinates (ρ, θ), the first fundamental form could be written as $d\rho^2 + G(r, \theta)d\theta^2$. We now show that, whenever we have local coordinates with respect to which the first fundamental form is of this shape, then we have a corresponding nice formula for the curvature. The reader should be surprised by the simplicity of this formula, which should be taken as further evidence for the importance of geodesic polar coordinates. The proof we give has been adapted from the treatment of the general case in Chapter 10 of [9]. The reason why the proof drops out reasonably cleanly is because we make our calculations with respect to a *moving frame*, which is well-suited to the problem. The reader should also note that the symmetry of mixed partial derivatives plays a crucial role in a couple of places. A special case of the result is that of surfaces of revolution, determined by a unit speed curve $\eta(u) = (f(u), 0, g(u))$, where we saw that the first fundamental form was $du^2 + f^2\, dv^2$ and the curvature was $-f''/f$ (Proposition 6.20).

Theorem 8.1 *Suppose S is an embedded surface with a smooth parametrization $\sigma : V \to U \subset S$, on which the first fundamental form takes the shape $du^2 + G(u, v)dv^2$. Then the Gaussian curvature is*

$$K = \frac{-(\sqrt{G})_{uu}}{\sqrt{G}}.$$

Proof   For any given point of V, we set $e = \sigma_u$ and $f = \sigma_v/\sqrt{G}$; together with $\mathbf{N}$, these form an orthonormal basis of $\mathbf{R}^3$ (using the assumption concerning the first fundamental form). As $e \cdot e = 1$, we have by differentiation that $e \cdot e_u = 0$. Thus (and similarly) we can write

$$
\begin{aligned}
e_u &= \alpha f + \lambda_1 \mathbf{N}, & e_v &= \beta f + \mu_1 \mathbf{N}, \\
f_u &= -\alpha' e + \lambda_2 \mathbf{N}, & f_v &= -\beta' e + \mu_2 \mathbf{N}.
\end{aligned}
\tag{$*$}
$$

Since $e \cdot f = 0$, we also have

$$e_u \cdot f + e \cdot f_u = 0,$$
$$e_v \cdot f + e \cdot f_v = 0.$$

Using these two equations, we deduce from $((*))$ that $\alpha' = \alpha$ and $\beta' = \beta$. But

$$\alpha = e_u \cdot f = \sigma_{uu} \cdot \sigma_v / \sqrt{G}$$
$$= (\sigma_u \cdot \sigma_v)_u / \sqrt{G} - \frac{1}{2}(\sigma_u \cdot \sigma_u)_v / \sqrt{G}$$
$$= 0 + 0,$$

and

$$\beta = e_v \cdot f = \sigma_{uv} \cdot \sigma_v / \sqrt{G}$$
$$= \frac{1}{2} G_u / \sqrt{G} = (\sqrt{G})_u.$$

Now, using $((*))$ again,

$$\lambda_1 \mu_2 - \lambda_2 \mu_1 = e_u \cdot f_v - f_u \cdot e_v$$
$$= \frac{\partial}{\partial u}(e \cdot f_v) - \frac{\partial}{\partial v}(e \cdot f_u)$$
$$= -\beta_u + 0$$
$$= -(\sqrt{G})_{uu}, \qquad\qquad (**)$$

using the fact that $e \cdot f_u = -\alpha = 0$, and the above formula for β.
We now use Proposition 6.17:

$$\mathbf{N}_u \times \mathbf{N}_v = (a\sigma_u + b\sigma_v) \times (c\sigma_u + d\sigma_v)$$
$$= (ad - bc)\,\sigma_u \times \sigma_v$$
$$= K\,\sigma_u \times \sigma_v,$$

where $\mathbf{N} = \sigma_u \times \sigma_v / \sqrt{G} = e \times f$. Therefore,

$$K\sqrt{G} = (\mathbf{N}_u \times \mathbf{N}_v) \cdot \mathbf{N} = (\mathbf{N}_u \times \mathbf{N}_v) \cdot (e \times f)$$
$$= (\mathbf{N}_u \cdot e)(\mathbf{N}_v \cdot f) - (\mathbf{N}_u \cdot f)(\mathbf{N}_v \cdot e)$$
$$= (\mathbf{N} \cdot e_u)(\mathbf{N} \cdot f_v) - (\mathbf{N} \cdot f_u)(\mathbf{N} \cdot e_v)$$

(noting that $\mathbf{N} \cdot e = 0$ implies that $\mathbf{N}_u \cdot e + \mathbf{N} \cdot e_u = 0$ and $\mathbf{N}_v \cdot e + \mathbf{N} \cdot e_v = 0$, with analogous formulae holding for f).

Calculating these dot products, we have

$$K\sqrt{G} = \lambda_1 \mu_2 - \lambda_2 \mu_1$$
$$= -(\sqrt{G})_{uu}$$

(the second equality coming from (∗∗)), and therefore we obtain the formula

$$K = -(\sqrt{G})_{uu}/\sqrt{G}. \qquad \square$$

For P a point on an embedded surface $S \subset \mathbf{R}^3$, the local geodesic polar coordinates (ρ, θ) at P, and hence also the function $G(\rho, \theta)$, depend only on the first fundamental form (i.e. the metric). The curvature on the corresponding coordinate patch on S is then given by $K = -(\sqrt{G})_{\rho\rho}/\sqrt{G}$. The point P corresponds to $\rho = 0$ and is technically not in the coordinate patch; in fact, we saw in Lemma 7.16 that $\lim_{\rho \to 0} G = 0$. However, all our functions, including the curvature, are smooth, and so the above equation also determines (in the limit as $\rho \to 0$) the curvature at P. Thus we deduce the following corollary, a result of which even Gauss was rather proud, although this is not the proof he gave. Gauss himself used the adjective egregium (which translates into *remarkable* or *outstanding*) to describe this result.

Corollary 8.2 (Gauss's Theorema Egregium)   *The curvature of an embedded surface depends only on the first fundamental form. In particular, if two embedded surfaces locally have isometric charts, then the curvatures are locally the same.* $\qquad \square$

8.2 Abstract smooth surfaces and isometries

In Chapter 4, we studied arbitrary Riemannian metrics on open subsets of $\mathbf{R}^2$. In Chapter 6, we studied embedded surfaces S in $\mathbf{R}^3$. These were covered by charts, identifying open subsets of S homeomorphically with open subsets of $\mathbf{R}^2$, and the identifications were consistent with the natural smooth structures on these open subsets of $\mathbf{R}^2$ by Proposition 6.2. However, these open subsets of $\mathbf{R}^2$ also carry Riemannian metrics corresponding to the first fundamental form (which is induced on the tangent spaces from the standard dot product on $\mathbf{R}^3$), and the above identifications are also consistent with these metrics. We now abstract these properties in a natural way in order to generalize the geometries from both chapters into one which subsumes them both, namely an *abstract surface* carrying a *Riemannian metric*.

Definition 8.3   An *abstract (smooth) surface* is a metric space S with homeomorphisms $\theta_i : U_i \to V_i$ from open subsets $U_i \subset S$ to open subsets $V_i \subset \mathbf{R}^2$ (i ranging over an indexing set I) such that

(i) $S = \bigcup_{i \in I} U_i$, and
(ii)   for $i, j \in I$, the map $\phi_{ij} := \theta_i \circ \theta_j^{-1} : \theta_j(U_i \cap U_j) \to \theta_i(U_i \cap U_j)$ is a diffeomorphism.
(iii) We shall assume also that the space is *connected*.

As in the embedded case, we call the θ_i *charts*, the collection of θ_i an *atlas*, and the ϕ_{ij} *transition functions* (cf. Definition 6.5). In what follows, we often just refer to S as a surface.

We say that S is *closed* if it is compact; e.g. by Bolzano–Weierstrass, an embedded surface $S \subset \mathbf{R}^3$ is compact if and only if it is closed in $\mathbf{R}^3$ and bounded.

Given a continuous curve $\gamma : [a, b] \to S$, we say that γ is *smooth* if, whenever $\gamma(t) \in U_i$ for some chart $\theta_i : U_i \to V_i$, the composite $\theta_i \circ \gamma$ is locally a smooth curve on V_i. Since, by definition, the transition functions are smooth, this condition does not depend on the choice of chart containing $\gamma(t)$.

For an abstract smooth surface S (equipped with an atlas), a *Riemannian metric* on S is defined to be given by *Riemannian metrics* on the images V_i of the charts $\theta_i : U_i \to V_i \subset \mathbf{R}^2$, subject to compatibility conditions that for all i, j (and $\phi = \phi_{ij}$)

$$\langle d\phi_P(\mathbf{a}), d\phi_P(\mathbf{b}) \rangle_{\phi(P)} = \langle \mathbf{a}, \mathbf{b} \rangle_P$$

inner-product given   inner-product given

by metric on V_i by metric on V_j

for all $P \in \theta_j(U_i \cap U_j)$ and $\mathbf{a}, \mathbf{b} \in \mathbf{R}^2$. This is just the statement that the transition functions are isometries of the Riemannian metrics on the open subsets of $\mathbf{R}^2$, in the sense of Chapter 4.

This enables us to define lengths and energies of curves on an abstract surface S, areas of regions on S, and geodesics on S merely by looking at the corresponding charts (this is entirely analogous to the case of embedded surfaces). The fact that these concepts are well defined follows from the invariance of length, energy and area under isometries; this was proved in Chapter 4 for lengths and areas, and the proof given there for lengths works equally well for energies.

Example   The three classical geometries we studied before are:

(i)   The Euclidean plane $\mathbf{R}^2$, with Riemannian metric $dx^2 + dy^2$.

(ii)   The embedded surface $S^2 \subset \mathbf{R}^3$, with metric induced from the Euclidean metric on $\mathbf{R}^3$.

(iii)   The unit disc model D of the hyperbolic plane with Riemannian metric $4(dx^2 + dy^2)/(1 - (x^2 + y^2))^2$, or equivalently the upper half-plane model H with Riemannian metric $(dx^2 + dy^2)/y^2$.

In the cases (i) and (iii), we only need one chart ($\theta = $ id) to define the abstract surface with its Riemannian metric. By a theorem of Hilbert, the hyperbolic plane cannot be realized as an embedded surface.

Example   We saw before that the torus $T \subset \mathbf{R}^3$ has natural charts $\theta : U \to V$, arising from the projection $\varphi : \mathbf{R}^2 \to T$, where U is the complement of two circles in T and V is the interior of a unit square in $\mathbf{R}^2$. In this case, the transition functions associated to an atlas consisting of such charts are *locally* just translations in $\mathbf{R}^2$, and so clearly satisfy the required isometry condition, with respect to the Euclidean metric on $\mathbf{R}^2$ — see however Exercise 8.1. From this, it follows that T is an abstract smooth surface, which comes equipped with the locally Euclidean Riemannian metric.

There are different Riemannian metrics which can be placed on T. One is the metric that arises from considering T as an embedded surface in $\mathbf{R}^3$, and this we encountered in Chapter 6 (where we found it convenient to study the embedded surface as a surface of revolution). The more natural metric on T is however the locally Euclidean metric. We observed in Chapter 6 that T with this metric cannot be realized as an embedded surface in $\mathbf{R}^3$. As T is compact, it is also not homeomorphic to an open subset of $\mathbf{R}^2$.

Given an abstract surface S equipped with a Riemannian metric, we can define an associated metric on S in the same way as Section 4.3 by

$$\rho_S(x_1, x_2) = \inf\{\text{length } \Gamma : \Gamma \text{ a piecewise smooth curve joining } x_1 \text{ and } x_2\}.$$

The implications of this are twofold. The usual definition of an abstract surface is in terms of a Hausdorff topological space rather than a metric space; in the presence of a Riemannian metric however, there is a natural metric on the space defined by the above recipe. Moreover, if we have taken our definition of an abstract surface in terms of a metric space, then the given metric may well not be the natural metric to consider. For instance, for an embedded surface $S \subset \mathbf{R}^3$, we have a metric on S given by taking distances between points in $\mathbf{R}^3$, but (assuming S to be connected) the natural metric to consider is the one induced from the Riemannian metric on S (as given by the first fundamental form).

Definition 8.4   A map $f : X \to Y$ between abstract surfaces is *smooth* if for any charts $\theta : U \to V$ on X and $\theta^* : U^* \to V^*$ on Y with $U \cap f^{-1}(U^*) \neq \emptyset$, the composite map

$$\bar{f} = \theta^* \circ f \circ \theta^{-1} : \theta(U \cap f^{-1}(U^*)) \subset V \to V^*$$

is smooth. This is saying that, once we have identified the domains of charts on X and Y as open subsets of $\mathbf{R}^2$ via the relevant charts, the induced map between the appropriate open subsets of $\mathbf{R}^2$ is smooth.

A smooth map f is called a *diffeomorphism* if it has a smooth inverse.

Suppose now X and Y have Riemannian metrics. A smooth map f is called a *local isometry* if for all pairs of charts as above, $\bar{f}$ preserves the Riemannian metric, that is, for all $P \in \theta(U \cap f^{-1}(U^*))$ and $\mathbf{a}, \mathbf{b} \in \mathbf{R}^2$,

$$\langle \mathbf{a}, \mathbf{b} \rangle_P = \langle d\bar{f}_P(\mathbf{a}), d\bar{f}_P(\mathbf{b}) \rangle^*_{\bar{f}(P)}.$$

$$\underset{\substack{\text{inner-product} \\ \text{for } \theta \text{ chart}}}{} \qquad \underset{\substack{\text{inner-product} \\ \text{for } \theta^* \text{ chart}}}{}$$

If f is also a diffeomorphism, then it is called an *isometry* — lengths and areas are then preserved under f. Moreover, an isometry will also preserve the associated intrinsic metrics we defined above, i.e. $\rho_Y(f(x_1), f(x_2)) = \rho_X(x_1, x_2)$ for all $x_1, x_2 \in X$.

Example

(i)   In Chapter 5, we defined an isometry between the upper half-plane and the disc models of the hyperbolic plane $H \to D$ — in fact, we defined the Riemannian metric on H in order that our given map was an isometry.

(ii)   We have the quotient map of surfaces $\varphi : \mathbf{R}^2 \to T$, where T is the torus.

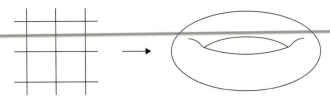

Taking the locally Euclidean metric on T and the Euclidean metric on $\mathbf{R}^2$, we know that φ is a local isometry, but it is clearly not an isometry.

(iii)   On a compact orientable surface S of genus g (i.e. a g-holed torus), one can prove the existence of a locally hyperbolic Riemannian metric on S, and also that there exists a local isometry $f : D \to S$ from the hyperbolic plane to S.

The theory from Chapter 7 extends immediately to any abstract smooth surface S equipped with a Riemannian metric, to yield a normal neighbourhood W for any point $P \in S$, and local geodesic polar coordinates (ρ, θ), with respect to which the metric takes the form $d\rho^2 + G(\rho, \theta)d\theta^2$; here $G(\rho, \theta)$ is a smooth function in the coordinates. Theorem 8.1 suggests that we should *define*, for all points of $W \setminus \{P\}$, the *Gaussian curvature* by the formula

$$K := -(\sqrt{G})_{\rho\rho}/\sqrt{G}.$$

By Theorem 8.1, this is consistent with our definition of the Gaussian curvature in the embedded case. It will prove to be slightly more convenient to impose a further condition on the normal neighbourhoods W we use, namely that they should be *strong normal neighbourhoods*.

Definition 8.5   A normal neighbourhood W for which any two points are connected by at most one geodesic in W will be called a *strong* normal neighbourhood. For instance, on the sphere, open balls of radius $\delta > 0$ are normal neighbourhoods when $\delta < \pi$, but are not strong normal neighbourhoods unless $\delta < \pi/2$. More generally, we shall see in Proposition 8.12 that sufficiently small normal neighbourhoods of a point are always strong normal neighbourhoods (implied by the definition of strongly convex in Section 8.4). The reader should perhaps be warned that the terminology introduced here is non-standard.

For the above to be a good definition of curvature, we need to prove two things:

(i)   For any $Q \in S$, we can find a strong normal neighbourhood W of some point $P \neq Q$ with $Q \in W$.

(ii)   The value of K does not depend on the particular choice of point P and strong normal neighbourhood W.

Here, we show that (i) holds, and we shall prove (ii) in the next section. Given a point $Q \in S$, we may by Proposition 8.12 choose a strong normal neighbourhood, a geodesic ball $B(Q, \delta)$ of radius $\delta > 0$. Let us consider the compact subset given by the closed ball $\bar{B} = \bar{B}(Q, \delta/2)$. For each $P \in \bar{B}$, there exists $0 < \varepsilon(P) < \delta/2$ such that the open ball $B(P, \varepsilon(P))$ is a normal neighbourhood of P; moreover, ε may be chosen to be a continuous function of P (Theorem 7.13). As such, it attains its bounds on $\bar{B}$ (Exercise 1.10), and so there exists $0 < \varepsilon_0 < \delta/2$ such that $B(P, \varepsilon_0)$ is a normal neighbourhood of P, for all $P \in \bar{B}$. Since $B(P, \varepsilon_0) \subset B(Q, \delta)$ for all P, these are in fact strong normal neighbourhoods. In particular, if we now choose any point $P \neq Q$ whose geodesic distance from Q is strictly less than ε_0, then Q is in the punctured strong normal neighbourhood $B(P, \varepsilon_0) \setminus \{P\}$, and so (i) has been proved.

Provided we show that our definition does not depend on the particular choice of geodesic polar coordinate system, a consequence of the above definition is that the curvature is a smooth function on S, since on any strong normal neighbourhood W of a point $P \in S$, it is clearly smooth on $W \setminus \{P\}$.

Before going on to prove (ii), let us consider our definition in the case of the three classical geometries. For $\mathbf{R}^2$ and S^2, we already know that our definition gives the correct answer for the curvature, as both are embedded surfaces. In the case of the disc model D of the hyperbolic plane, for any given point $P \in D$, we can find an isometry of D which sends P to the origin. Thus, the calculation we perform in (iii) below verifies that the curvature is constant with value -1.

Example

(i) For $\mathbf{R}^2$, we have $\rho = r$, and the metric is $d\rho^2 + \rho^2 d\theta^2$; thus $\sqrt{G} = \rho$ and $K = 0$.

(ii) For S^2, the metric is $d\rho^2 + \sin^2\rho\, d\theta^2$ (with respect to local geodesic polar coordinates). Therefore $\sqrt{G} = \sin\rho$, and so $(\sqrt{G})_{\rho\rho} = -\sqrt{G}$ and $K = 1$.

(iii) On the disc model of the hyperbolic plane, the Riemannian metric is given (in geodesic polar coordinates) by $d\rho^2 + \sinh^2\rho\, d\theta^2$. Therefore $\sqrt{G} = \sinh\rho$, and so $K = -1$.

8.3 Gauss–Bonnet for geodesic triangles

Let S denote an abstract smooth surface equipped with a Riemannian metric. For $R \subset S$ a suitably well-behaved region and K a continuous function on R, we are able to form an integral $\int_R K\, dA$. This should be understood on appropriate charts with coordinates (u, v) — in the usual notation, it is just $\int K\, (EG - F^2)^{1/2}\, du\, dv$, the integral being taken over the appropriate region in $\mathbf{R}^2$. For a region contained in the domains of two different charts on S, the transition function is (by definition) an isometry with respect to the Riemannian metrics, and the fact that the integral is well defined follows as in the proof of Proposition 4.7. This enables us to define the integral, even if R is not contained in the domain of a single chart. When $K = 1$, we just recover the area of R.

Regions over which we shall be particularly interested in integrating are (geodesic) polygons on S, defined precisely as in Definition 3.4 in the case of S^2 and T. An even more particular case of interest for us will be the case of *geodesic triangles*, whose sides we shall assume (as in the spherical case) to have the property that they are the unique curves of absolute minimum length joining the relevant vertices. In the spherical case, this definition would however include the case of the complement of a spherical triangle — we shall therefore usually impose a further condition on our geodesic triangles that they be contained in a strong normal neighbourhood of one of their vertices.

The angles of such a geodesic triangle are determined via the Riemannian metric on any suitable chart, by taking the inner-product of tangents to the sides (we note below that the angles are non-reflex); because of the compatibility conditions that are stipulated in the definition of a Riemannian metric on an abstract surface, this will not depend on the choice of chart.

Lemma 8.6   *Suppose that W is a strong normal neighbourhood of $A \in S$, and B, C are distinct points of $W \setminus \{A\}$, such that the curve Γ of absolute minimum length joining B to C lies in $W \setminus \{A\}$. The vertices A, B, C then determine a (unique) geodesic triangle in W.*

Proof   Observe that for each point P on Γ, the (unique) geodesic ray from A and passing though P intersects Γ only at P. If it were to intersect Γ at another point $Q \neq P$, then there would be two geodesics in W joining P to Q (since clearly the radial segment PQ cannot form part of Γ), contradicting the assumption that W was a *strong* normal neighbourhood. Now W is the image of some open ball $B_\varepsilon = B(0, \varepsilon)$ under the parametrization σ, obtained via normal coordinates as in Theorem 7.13, where the geodesic rays from A now correspond in B_ε to the radial rays from the origin. The geodesic curve in B_ε corresponding to Γ in W is then a curve, contained in the sector of angle α ($< \pi$) in $B(0, \varepsilon)$ determined by the radial rays corresponding to AB and AC. This last statement follows from W being a *strong* normal neighbourhood: if the curve were contained in the complementary sector with reflex angle $2\pi - \alpha$, then it would intersect the *diameter* corresponding to AB twice (once on either side of the origin), and hence it and the diameter would be different geodesics in B_ε joining the same two points.

The radial geodesic segments AB and AC are absolutely length minimizing by Corollary 7.18, as by assumption is the geodesic segment BC. The concatenation of the three geodesic segments AB, BC and CA in W corresponds to a simple closed curve in B_ε of a particularly accessible type — in particular, we know that its complement in $\mathbf{R}^2$ has precisely two connected components, with the bounded one being contained in B_ε (see Exercise 1.13). Moreover, the closure of this bounded component is the union of initial segments of radial rays, with arguments in a range $[\theta_0, \theta_0 + \alpha]$, where α is the angle of the geodesic triangle at A — the image under σ of this set is the geodesic triangle in W we seek. The above description ensures that all three angles in the triangle are non-reflex (a fact we quoted above).   □

We assume now that a geodesic triangle $\triangle = ABC$ is contained in a strong normal neighbourhood W of one of its vertices, say A. By considering geodesic polar coordinates (r, θ) on W, the Riemannian metric then takes the form $dr^2 + G(r, \theta)d\theta^2$

on the punctured neighbourhood. We may define a curvature function K on $W \setminus \{A\}$ by the formula from the previous section, namely $K := -(\sqrt{G})_{\rho\rho}/\sqrt{G}$.

Proposition 8.7 *Suppose that a geodesic triangle $\triangle$ is contained in a strong normal neighbourhood of one of its vertices, and has internal angles α, β and γ. If the curvature function K is defined as above on the punctured normal neighbourhood, then*

$$\int_{\triangle} K \, dA = (\alpha + \beta + \gamma) - \pi.$$

Proof   The most direct proof of this is by explicitly integrating, in an analogous way to our proof in the case of a hyperbolic triangle. We shall take geodesic polar coordinates (r, θ) on the strong normal neighbourhood, and denote by $\sigma : V \to U$ the corresponding parametrization of an open subset $U \subset S$, with V an open subset of $\mathbf{R}^2$ of the form $(0, \delta) \times (\lambda, \lambda + 2\pi)$. The sides of the triangle containing A correspond to taking constant values of the θ coordinate, which we may assume to be $\theta = 0$ and $\theta = \alpha$, where we assume λ has been chosen so that $[0, \alpha] \subset (\lambda, \lambda + 2\pi)$. There is a very minor objection to taking such a parametrization by geodesic polars, in that its image does not include the vertex A of our triangle, but this may be got around by a suitable limiting argument, as the integrations we perform will not be affected by omitting the vertex A. The corresponding Riemannian metric $\langle \, , \, \rangle$ on V is then just $dr^2 + G(r, \theta) \, d\theta^2$.

We observed above that, for each point P on the third side Γ of the geodesic triangle, namely the side BC, the (unique) geodesic ray from A and passing through P intersects Γ only at P, and that the triangle is just the union of such geodesic segments from A to a point P of Γ.

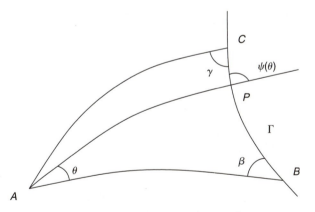

We deduce therefore that Γ may be parametrized by the coordinate θ, with $\Gamma(\theta) = \sigma(f(\theta), \theta)$ representing the unique point of intersection of the geodesic ray γ_θ with Γ, where $0 \leq \theta \leq \alpha$ (with $B = \Gamma(0)$ and $C = \Gamma(\alpha)$). Here, the function $f(\theta)$ is just the length of the radial geodesic from A to the point $P = \Gamma(\theta)$. We let $\psi(\theta)$ denote the angle, as shown in the diagram, at which the curve Γ meets the radial geodesic corresponding to a given value of θ; in particular, $\psi(0) = \pi - \beta$ and $\psi(\alpha) = \gamma$.

We now work entirely on the chart V. The three sides of the triangle are then given by the rays $\theta = 0$ and $\theta = \alpha$, and the curve $\eta(\theta) = (f(\theta), \theta)$. If we let s denote the arc-length of η, then $ds/d\theta = \|\eta'(\theta)\| = h(\theta)$ say, where $h(\theta) = \left(f'(\theta)^2 + G(f(\theta), \theta)\right)^{1/2} > 0$. Here, as in the rest of the proof, we shall use the prime notation for differentiation with respect to the parameter θ. If we reparametrize η in terms of its arc-length, then it satisfies the geodesic equations (7.3); the first of these is $2\, d^2 f/ds^2 = G_r (d\theta/ds)^2$, or in terms of θ-derivatives that

$$\frac{1}{h}\left(\frac{1}{h}f'\right)' = \frac{1}{2h^2}G_r.$$

We know that e_1 and $e_2/\sqrt{G}$ form an orthonormal basis for $\mathbf{R}^2$ with respect to the metric, and that $\eta'(\theta) = (f'(\theta), 1)$ has norm $h(\theta)$. We may identify the angle ψ by means of the relation $\cos\psi = \langle e_1, \eta'\rangle/h = f'/h$. We also have the relation

$$h\sqrt{G}\sin\psi = \langle e_2, \eta'\rangle = \|e_2\|^2 = G,$$

that is $\sin\psi = \sqrt{G}/h$. Differentiating the first of these relations with respect to s gives $-\psi'\sin\psi/h = (f'/h)'/h = G_r/2h^2$, using the above geodesic condition, and therefore (using the second of the relations to substitute for $\sin\psi$) that

$$\psi' = -\frac{1}{2}G_r/\sqrt{G} = -(\sqrt{G})_r.$$

Using the formula $K = -(\sqrt{G})_{rr}/\sqrt{G}$ for the curvature function, the integral we want is

$$\int_\triangle K\, dA = \int_0^\alpha \int_0^{f(\theta)} K\sqrt{G}\, dr\, d\theta = -\int_0^\alpha \int_0^{f(\theta)} (\sqrt{G})_{rr}\, dr\, d\theta.$$

We integrate with respect to r, and use the relation $\psi'(\theta) = -(\sqrt{G})_r(f(\theta), \theta)$ and the fact (see Remark 7.17) that $(\sqrt{G})_r \to 1$ as $r \to 0$, obtaining

$$\int_\triangle K\, dA = \int_0^\alpha (\psi' + 1)\, d\theta = \gamma - (\pi - \beta) + \alpha,$$

as required.   □

We can now show that the curvature takes a well-defined value, independent of choices made — in doing so, we shall recover an equivalent, but far more geometric, definition for the curvature in Definition 8.8.

Suppose that W is a strong normal neighbourhood of some point $P \in S$. With respect to the geodesic polar coordinates (r, θ) on W, the Riemannian metric then takes the form $dr^2 + G(r, \theta)d\theta^2$ on $W \setminus \{P\}$, and we define a smooth function K on $W \setminus \{P\}$ by $K := -(\sqrt{G})_{\rho\rho}/\sqrt{G}$.

Let us take any point $Q \in W \setminus \{P\}$. We shall show in Lemma 8.13 that, for sufficiently small $\varepsilon > 0$, the geodesic ball U with centre Q and radius ε has the property that any three distinct points A, B, C of U determine a unique geodesic triangle in U; we assume also that ε is chosen with $U \subset W \setminus \{P\}$. Moreover, such a triangle may be expressed in terms of unions and set-theoretic differences of the three geodesic triangles PAB, PBC and PCA in W, as shown in the diagram below (the third case illustrates the possibility that PBC may degenerate, and so only two triangles are involved). The main reason for having taken a *strong* normal neighbourhood W was to ensure that these triangles exist (Lemma 8.6).

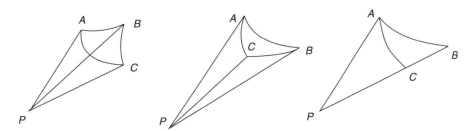

We can apply Proposition 8.7 to each of the geodesic triangles PAB, PBC and PCA; by taking appropriate sums and differences of the resulting formulae, we deduce that $\int_{\triangle} K \, dA = (\alpha + \beta + \gamma) - \pi$, where α, β and γ denote the internal angles of $\triangle$.

If now U is a neighbourhood of Q of the above type, contained in two different punctured strong normal neighourhoods $W_1 \setminus \{P_1\}$ and $W_2 \setminus \{P_2\}$ of P_1 (respectively P_2), and K_1 (respectively K_2) denote the corresponding curvature functions on U, then the above argument shows that, for any triangle $\triangle \subset U$,

$$\int_{\triangle} K_1 \, dA = \int_{\triangle} K_2 \, dA.$$

If we now choose smaller and smaller triangles $\triangle \subset U$ which *contain* the point Q (for instance, with Q a vertex), we may deduce that $K_1(Q) = K_2(Q)$. This was the second claim made in the previous section, that the curvature we had defined did not depend on the choice of geodesic polar coordinate system.

Moreover, we see that the somewhat opaque definition of curvature we made above is equivalent to a very attractive geometric definition of curvature, namely:

Definition 8.8 The curvature K at any point $Q \in S$ may be recovered from the geodesic triangles $\triangle$ of small diameter containing Q, by the formula

$$K = \lim_{\text{diam } \triangle \to 0} \left(\frac{\sum \text{angles of } \triangle - \pi}{\text{area } \triangle} \right).$$

Having shown now that the curvature function is both well defined and well behaved, we have other equivalent definitions, which provide further insight. Letting P denote a point on a smooth abstract surface S, equipped with a Riemannian metric, we have local geodesic polar coordinates (ρ, θ), with respect to which the metric is $d\rho^2 + G(\rho, \theta) \, d\theta^2$. By Lemma 7.16, $\sqrt{G(\rho, \theta)}$ is a smooth function of ρ and θ

(see also Remark 7.17); assuming for simplicity that it may be expanded as a power series in ρ with coefficients depending on θ (the general case is a straightforward extension, which is left to the reader), the facts from Remark 7.17 that $\sqrt{G} \to 0$ and $(\sqrt{G})_\rho \to 1$ as $\rho \to 0$, together with the characterization of the curvature at P as $K = \lim_{\rho \to 0}(\sqrt{G})_{\rho\rho}/\sqrt{G}$, ensure that $\sqrt{G}$ is locally of the form

$$\sqrt{G(\rho,\theta)} = \rho\,(1 - K\rho^2/6 + \text{ higher order terms in } \rho\,).$$

Note that we have managed to find two further terms in the expansion of $\sqrt{G}$, as compared to Remark 7.17.

If we now take a small geodesic circle of radius ε with centre at P, with circumference $C(\varepsilon)$ and area $A(\varepsilon)$, one checks easily that we may recover the curvature K as the limits as $\varepsilon \to 0$ of both

$$3(2\pi\varepsilon - C(\varepsilon))/\pi\varepsilon^3 \quad \text{and} \quad 12(\pi\varepsilon^2 - A(\varepsilon))/\pi\varepsilon^4.$$

The geometric characterizations of curvature which we have produced in the last few paragraphs are generalizations of calculations we made in the specific cases of the sphere and hyperbolic plane.

Remark   With the curvature K now shown to be well defined, Proposition 8.7 is just the Gauss–Bonnet formula for geodesic triangles of the form being considered. This will be a crucial ingredient in our proof of the global Gauss–Bonnet theorem, which we prove in the following section.

Finally in this section, we mention a couple more basic results on curvature. The first of these describes how the curvature behaves under rescaling the metric by a positive real number c^2; for instance, a sphere of radius $c > 0$ has curvature $1/c^2$.

Lemma 8.9   *Suppose S is a surface equipped with a Riemannian metric g with curvature K. If we consider the scaled metric c^2g on S, then the curvature becomes K/c^2.*

Proof   This follows most easily from our definition of curvature in terms of small geodesic triangles containing a given point Q. Note that scaling the metric does not change angles, and so for a given small triangle $\triangle$ containing Q, we have that $\int_\triangle K\, dA$ is invariant under the scaling. Since the area of $\triangle$ scales by c^2, the claim follows directly from the above definition. $\square$

Suppose we have an abstract smooth surface equipped with a Riemannian metric, which is locally isometric to the spherical metric, the Euclidean metric or the hyperbolic metric. This then ensures that the Gaussian curvature on S is constant, namely 1, 0 or -1 respectively. We can now prove the converse to this, a classical result due to Minding, that a surface with a constant curvature metric is (after rescaling the metric) locally isometric to an open subset of one of the three basic classical geometries.

Theorem 8.10 *If S is a smooth surface equipped with a Riemannian metric g with constant curvature K, then after a suitable (constant) rescaling of the metric, the surface S is locally isometric to an open subset of S^2, $\mathbf{R}^2$ or the hyperbolic plane (according to whether $K > 0$, $K = 0$ or $K < 0$).*

Proof We may rescale the metric, and so by Lemma 8.9, we may assume that the curvature $K = 1, 0$ or -1. We may write the metric in local geodesic polar coodinates

$$d\rho^2 + G(\rho, \theta)d\theta^2.$$

We saw in Remark 7.17 that $G \to 0$ and $(\sqrt{G})_\rho \to 1$ as $\rho \to 0$.

Let us take a fixed value for θ and set $f(\rho) = \sqrt{G(\rho, \theta)}$. From the formula for curvature, we know that the function $f(\rho)$ satisfies the differential equation

$$f_{\rho\rho} + Kf = 0.$$

Given that $f \to 0$ and $f_\rho \to 1$ as $\rho \to 0$, we deduce in the three cases that $f = \sin \rho$, $f = \rho$ and $f = \sinh \rho$. From this, it is an easy exercise (Exercise 8.4) to show that there exists locally an isometry to an open subset of the appropriate standard model above. $\square$

8.4 Gauss–Bonnet for general closed surfaces

Let S be a compact abstract smooth surface, equipped with a Riemannian metric. We may define (as we did in Chapter 3 for the sphere and torus) the Euler number of a triangulation on S as $e = F - E + V$, where $F = $ # faces, $E = $ # edges and $V = $ # vertices. We shall show that this is a topological invariant of the surface in an almost identical way to that we used for the sphere and torus, by replacing any topological triangulation of S by a polygonal decomposition with the same Euler number, and by proving a generalized global version of the Gauss–Bonnet theorem.

If we start from just a compact abstract smooth surface S, then usually the existence of both a Riemannian metric and a triangulation on S will be straightforward. For instance, an embedded g-holed torus in $\mathbf{R}^3$ carries an obvious Riemannian metric, and we argued in Chapter 3 that there exists a triangulation. The existence in general of a Riemannian metric on S in fact follows by an easy argument, which simply patches together local Riemannian metrics on charts ([6], Lemma 2.3.3 or [12], page 309). The existence in general of a triangulation on S is perhaps most easily proved by using such a metric, and the convexity arguments from this section, in order to produce a *geodesic* triangulation. One proves first that there is a (geodesic) polygonal decomposition of S (see [6], Theorem 2.3.A.1), from which by further decomposition we may obtain a geodesic triangulation.

The proofs given earlier in the case of the sphere and torus for the global Gauss–Bonnet theorem and the topological invariance of the Euler number (in particular the proof of Theorem 2.16 and the detailed arguments from the appendix to Chapter 3) apply essentially unchanged to the general case of a compact surface S with a

Riemannian metric, once we have shown the existence of suitable convex open neighbourhoods. We shall therefore first need to say something about convexity.

In Chapters 2 and 3, we defined what it meant for a subset A of the sphere or locally Euclidean torus to be *convex*. This definition may be generalized in an obvious way to subsets B of a general surface S.

Definition 8.11   Let S denote an abstract smooth surface, equipped with a Riemannian metric. A subset B of S is called *convex* if, for any $Q_1, Q_2 \in B$, there is a unique length minimizing geodesic in S joining Q_1 to Q_2, and this curve is contained in B. A subset B of S is called *strongly convex* if it is convex, and for any points $Q_1, Q_2 \in B$, the length minimizing geodesic is the only geodesic in B joining the two points.

For example, the open balls of radius strictly less than $\pi/2$ for the sphere and $1/4$ for the locally Euclidean torus are easily seen to be strongly convex. We now prove the existence of strongly convex open neighbourhoods for any surface S equipped with a Riemannian metric. We comment that the crucial part of this argument appeared already at the end of Chapter 7, in Lemma 7.19.

Proposition 8.12   *Let P be any point on an abstract surface S equipped with a Riemannian metric. For all $\varepsilon > 0$ sufficiently small, we have that the following two properties hold.*

(i)   *The open geodesic ball $B(P, \varepsilon)$ is strongly convex, and*
(ii)   *for any $Q \in B(P, \varepsilon)$, the open geodesic ball $B(Q, 2\varepsilon)$ is a strong normal neighbourhood of Q.*

Proof   We choose a normal neighbourhood $W_0 = \sigma(B_\delta)$ around P, as defined in Theorem 7.13, so that the metric takes the form $dr^2 + G(r, \theta)\, d\theta^2$ with respect to geodesic polar coordinates (r, θ); we assume furthermore that δ has been chosen so that $G_r > 0$ on $B_\delta \setminus \{0\}$ (possible, since by Remark 7.17 we have $G_r/r \to 2$ as $r \to 0$). We consider the closed geodesic ball $\sigma(\bar{B}_{\delta/2})$, consisting of the points of S whose geodesic distance from P is at most $\delta/2$. By Theorem 7.13 and the argument we used to prove property (i) after the definition of Gaussian curvature in Section 8.2, there exists ε with $0 < 2\varepsilon < \delta/2$ such that, for all $Q \in \sigma(\bar{B}_{\delta/2})$, the geodesic ball centred on Q with radius 2ε is a normal neighbourhood of Q, contained in $W_0 = \sigma(B_\delta)$.

We show that the normal neighbourhood $W = \sigma(B_\varepsilon)$ of P is convex. Suppose we have points $Q_1, Q_2 \in W$; then

$$\rho(Q_1, Q_2) \le \rho(Q_1, P) + \rho(P, Q_2) < 2\varepsilon.$$

Considering the geodesic ball U of radius 2ε centred on Q_1, this contains a unique geodesic $\gamma : [0, 1] \to U$ of minimum length from Q_1 to Q_2, namely the radial geodesic (Corollary 7.18). Since any curve from Q_1 to a point on the boundary of $\bar{U}$ has length at least 2ε, this shows that γ is also the curve in S of absolute minimum length joining the two points. Note that U is contained in W_0.

If now the above curve γ contains P, it is a radial geodesic in W. Assume therefore that γ does not contain P; our assumptions imply that $G_r > 0$ at all points of γ. Since $\rho(P, Q_1) < \varepsilon$ and $\rho(P, Q_2) < \varepsilon$, we deduce from Lemma 7.19 that $\rho(P, \gamma(t)) < \varepsilon$ for $0 \le t \le 1$, and hence the curve γ is always contained in W; hence W is *convex*. Since however W is contained in the normal neighbourhood $U = B(Q_1, 2\varepsilon)$, in which the radial geodesic γ is the unique geodesic joining Q_1 to Q_2, we deduce that it is the unique geodesic joining Q_1 to Q_2 in W; hence W is *strongly* convex.

If now we repeat this argument, starting instead from a strongly convex normal neighbourhood $W_0 = \sigma(B_\delta)$ of P (just shown to exist), we obtain a strongly convex neighbourhood $W = \sigma(B_\varepsilon)$ of P, with the additional property that, for any $Q \in W$, the open geodesic ball $B(Q, 2\varepsilon)$ is a strong normal neighbourhood of Q. $\qquad\square$

We shall be arguing below with polygons contained in such strongly convex balls, and we shall need the convexity of any geodesic triangle contained in such a ball.

Lemma 8.13 *If $W = B(P, \varepsilon)$ is as in Proposition 8.12, then any three distinct points of W determine a unique geodesic triangle $\triangle \subset W$, and $\triangle$ is itself strongly convex.*

Proof   If the three points are A, B and C, we choose one of them, say A. The open geodesic ball $B(A, 2\varepsilon)$ is then a strong normal neighbourhood of A, and contains W. The minimum length geodesic Γ joining B to C is in the strongly convex open set W, and hence in $B(A, 2\varepsilon)$. We saw in the proof of Lemma 8.6 that the three points determine a unique geodesic triangle $\triangle \subset B(A, 2\varepsilon)$, which is moreover contained in the sector with angle $\alpha < \pi$ determined by the geodesic rays containing AB and AC. The description given there for $\triangle$ in terms of a union of geodesic rays from A, together with the convexity of W, ensures that $\triangle \subset W$.

Suppose now that we have two distinct points P, Q of $\triangle$; the minimum length geodesic γ from P to Q is contained in W, by convexity. We apply the argument from Lemma 8.6 again, deducing that γ must remain in the sector of $B(A, 2\varepsilon)$ determined by the geodesic rays containing AB and AC (since otherwise γ would intersect a diameter in more than one point). If therefore γ does not remain in $\triangle$, it must cross the third side Γ of $\triangle$ in more than one point, contradicting the *strong* convexity of W. Hence $\triangle$ is convex, and therefore automatically strongly convex (as it is contained in W). $\qquad\square$

The Gauss–Bonnet theorem for geodesic triangles contained in such a strongly convex open ball on S follows immediately from Proposition 8.7, since any such triangle is contained in a strong normal neighbourhood of any of its vertices. This now generalizes to geodesic polygons.

Corollary 8.14 *If Π is a geodesic n-gon on an abstract smooth surface S equipped with a Riemannian metric, and Π is contained in a strongly convex open ball W of* *the type constructed in Proposition 8.12, then*

$$\int_\Pi K \, dA = \sum_i \alpha_i - (n-2)\pi,$$

where $\alpha_1, \ldots, \alpha_n$ are the internal angles of Π.

Proof   This now follows from the case of geodesic triangles, and the induction argument we used for spherical polygons contained in a hemisphere. To start that argument, we needed to find a locally convex vertex P_2; again this is found by taking a point of the polygon at maximum distance from the centre P of the ball $W = \sigma(B_\varepsilon)$; this point is a vertex, since W was chosen with the property that, for any geodesic segment contained in W, the maximum distance from P occurs at an end-point (this follows from Lemma 7.19, given that G_r was assumed to be positive on $W \setminus \{P\}$).

If we take points Z_1 and Z_2 on the boundary of Π, either side of P_2 and sufficiently close to it, the geodesic triangle $\triangle = Z_1 P_2 Z_2$ will either be contained in Π (the case when P_2 is a locally convex vertex), or will have its interior disjoint from Π. We note in passing that Lemma 7.19 implies that, for such points Z_1 and Z_2, the points of $\triangle$ have distance from P at most $\rho(P, P_2)$. Moreover, in the second case, points sufficiently close to P_2 but not in $\triangle$ will lie in Π. If we consider points Q on the geodesic ray PP_2 just beyond P_2, these will be points of Π with $\rho(P, Q) > \rho(P, P_2)$, contradicting our initial choice of P_2.

We note that Π is also contained in a strong normal neighbourhood of any of its vertices, namely the geodesic ball of radius 2ε with centre at the vertex. Apart from standard convexity properties, the rest of the proof of Theorem 2.16 only used the properties of geodesics in the hemisphere that distinct geodesics meet in at most one point, and they have distinct tangents at any point of intersection; otherwise the proof was purely combinatorial. These two properties hold also in the case being considered, the second fact following from the uniqueness clause in Proposition 7.10; as our assumptions ensure that we have the convexity properties needed, including in particular Lemma 8.13, the proof of Theorem 2.16 therefore applies in the general case. We are thus able to express our polygon as the union of two simpler polygons meeting along a common side, for both of which we may assume that the required formula holds, and hence the general formula follows by induction.   □

Using this last result, we can now prove the global Gauss–Bonnet theorem, in an exactly analogous way to the method used for the sphere and torus in Chapter 3. In summary, we subdivide the triangulation so that each topological triangle is contained in a suitable strongly convex open set, replace the resulting triangulation by a polygonal decomposition of the surface, and then use the Gauss–Bonnet formula we have just proved for geodesic polygons.

Theorem 8.15 (Gauss–Bonnet theorem)   *Suppose S is a closed (i.e. compact) surface equipped with a Riemannian metric. Assuming the existence of a triangulation on S,* *we have*

$$\int_S K \, dA = 2\pi e,$$

where e is the Euler number. In particular, the Euler number depends on neither the choice of triangulation nor the choice of Riemannian metric, and so may be written as e(S).

Proof   The surface has a cover by open geodesic balls $B(P, \varepsilon(P)/2)$, where for each $P \in S$, the open ball $B(P, \varepsilon(P))$ is a strongly convex open ball of the type constructed in Proposition 8.12. Using compactness, we choose a finite subcover, and let ε be the minimum of the finite set of numbers $\varepsilon(P)$ occurring. If now Δ is any subset of S of diameter less than $\varepsilon/2$, then it must be contained in one of the corresponding finitely many strongly convex geodesic balls $B(P, \varepsilon(P))$.

Given then any topological triangulation of S, we can subdivide it using Construction 3.9, without changing the Euler number, so that each topological triangle has diameter less than $\varepsilon/2$, and hence is contained in a strongly convex open ball W of radius ε, of the type constructed by Proposition 8.12. We now use Construction 3.15 to polygonally approximate the edges of this triangulation by simple polygonal curves, and then arguing as in Proposition 3.16, we see that this yields a polygonal decomposition of S. We shall need the fact that the complement of a simple closed polygonal curve in S has at most two components, but since our geodesics (in appropriate geodesic polar coordinates) correspond to radial lines, the argument of Proposition 1.17 still applies (Remark 1.18), proving this fact.

Therefore, we have replaced the triangulation by a polygonal decomposition with the same Euler number, with each of the polygons being contained in a strongly convex open ball of the type constructed in Proposition 8.12. Hence, the Gauss–Bonnet formula holds for the integral of the curvature over each of these polygons, by Corollary 8.14. This then implies the required result, since by the argument from Proposition 3.13,

$$\sum_{n \geq 3} \sum_{n\text{-gons}} \left(\sum \text{interior angles} - (n-2)\pi \right) = 2\pi e.$$

$\square$

Example   The torus T with locally Euclidean metric clearly has K identically zero, and so $e(T) = 0$. If however we take the metric on T obtained from considering it as an embedded surface, then we saw in Chapter 6 that the curvature K takes positive, negative and zero values; nonetheless, Theorem 8.15 still says that $\int_T K \, dA = 0$ (see Exercises 8.2 and 8.3).

The global Gauss–Bonnet theorem implies directly the topological invariance of the Euler number. We suppose that S is a compact smooth surface, equipped with a Riemannian metric (the particular choice of Riemannian metric being irrelevant).

Corollary 8.16   *If X is a metric (or topological) space which is homeomorphic to S, then any (topological) triangulation on X has Euler number e(S).*

Proof   A topological triangulation on X gives rise, via the homeomorphism, to one with the same Euler number on S. This Euler number is however just $e(S)$, by the global Gauss–Bonnet theorem on S. $\square$

Remark 8.17   If one is prepared to restrict attention to compact *orientable* surfaces, and to triangulations of the surface whose edges are piecewise *smooth* curves, then there is another proof of the global Gauss–Bonnet theorem in terms of integrating the *geodesic curvature* round the edges of each triangle (Section 4.5 of [5], Chapter 12 of [8], or Chapter 11 of [9]). This proof does not yield the topological invariance of the Euler number, which would therefore need to be proved separately, for instance by the theory of homology groups from elementary Algebraic Topology.

8.5   Plumbing joints and building blocks

The fact that integrating the curvature over a closed surface gave such a basic invariant suggests that we might try also integrating the curvature over smooth *open* surfaces — this is the standard terminology for *non-compact* surfaces. There are many examples of smooth open surfaces where the area is infinite, but the curvature decays sufficiently rapidly for the integral to be finite — see for instance Exercises 8.7 and 8.8. In this section, we shall be interested in gluing together open surfaces to obtain a *compact* surface, and so the area will always be finite. We have already considered the integral of the curvature in Chapter 3 for the open hemisphere, where the answer was clearly 2π, from which we deduced that the real projective plane had Euler number 1. It will be slightly more convenient to draw out the equator into a *cylindrical end* or *neck* as shown below, where we shall take this neck to be a segment of a circular cylinder of radius 1 say. In order to do this in a smooth way, we need to modify the hemisphere in a neighbourhood of its boundary, but the reader will not doubt that this can be done, with the resulting surface S_0 being an embedded surface. We recall that the metric on a cylinder is locally Euclidean, with curvature therefore being zero.

S_0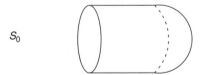

If now we have two copies S_0 and S_0' of this surface, we can glue them together along the cylindrical ends to achieve a smooth surface S, also an embedded surface (the metrics on the two cylindrical necks are the same). It is clear that S is just a deformed sphere, and hence has Euler number 2. The integral of the curvature over S is 4π by Gauss–Bonnet, and so the integral of the curvature over S_0 is 2π. The obvious additivity of the integral of the curvature will enable us to understand geometrically the general case of the g-holed torus, for $g > 0$.

The other basic *piece of plumbing* we shall need in order to construct the general g-holed torus is what the topologists termed a *pair of pants*: this surface also plays an important role in physics, in conformal field theory.

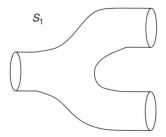

It is an embedded surface S_1 in $\mathbf{R}^3$ with three cylindrical ends, which we shall assume again to be segments of circular cylinders of radius one (within the properties stipulated, there is much flexibility about which surface we choose, but the choice here will not matter). To calculate the integral of the curvature over S_1, we can argue as follows: If we cap off the three cylindrical ends with copies of S_0, we obtain an embedded surface which is topologically the sphere. Since the integral of the curvature over this closed surface is 4π, and the integral of the curvature over each of the three caps S_0 is 2π, we deduce that the integral of the curvature over S_1 has to be -2π. The integral of the curvature divided by 2π should be thought as giving us the correct contribution to the Euler number, and may therefore be regarded as a *virtual Euler number*.

Example   We can form a surface which is topologically a torus by plumbing together two copies of S_1 in the way illustrated below to obtain a surface S_2, and then capping off the two remaining ends with copies of S_0. The Euler number we obtain is the sum of the virtual Euler numbers of the pieces, two of which are $+1$ and two of which are -1, giving a total of 0 as expected. We observe that the open surface S_2 constructed here has virtual Euler number -2.

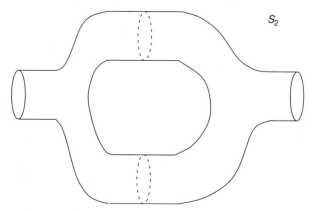

In general, we can form a g-holed torus, for $g > 0$, by gluing together g copies of S_2 in the obvious way, and capping off the two free ends by copies of S_0. Thus the

Euler number obtained is $2 - 2g$, this coinciding with the calculation we performed in Chapter 3 by means of triangulations.

There is another, at first sight more exotic, way in which we can understand the topology of the g-holed torus, in terms of children's building blocks. Let us consider the surface of a unit cube embedded in $\mathbf{R}^3$, and round off the edges and corners to achieve a smooth embedded surface S, which is homeomorphic to the original cube. Away from the vertices, we can round off an edge so that it looks locally like the product of a small arc of a smoothly embedded plane curve of unit speed with an open real interval, which as an embedded surface has a locally Euclidean first fundamental form (see Exercise 6.2). Geometrically, if we slice the cube by two suitable planes parallel to two given opposite faces, and take that region of the cube between these two planes, then for sufficiently close approximations S to the cube, the corresponding region of S may be described by a strip of paper bent appropriately, to form a surface which is the product of a 'rounded square' with an open real interval; the metric then corresponds to the locally Euclidean metric on the flat strip of paper. The reader is invited to convince herself that such smooth approximations to the unit cube exist.

The resulting metric on S is therefore locally Euclidean, apart from at points near where the vertices have been rounded off. When we integrate the curvature over S therefore, we only get contributions from these eight small neighbourhoods, each of which must therefore contribute $\pi/2$ to the integral. If we take the surface S to be a closer and closer approximation to the cube, the curvature concentrates in smaller and smaller such neighbourhoods. In the limit, we can think of the metric as being locally Euclidean on the surface of the cube minus the eight vertices, but that the curvature is now concentrated at the eight vertices. The contribution of each vertex to the Euler number is then $1/4$. This idea of curvature concentrating at points when we take limits is a common and fruitful one in more advanced differential geometry.

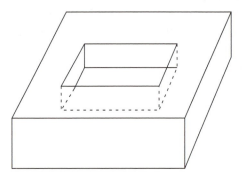

Suppose now our children's box of bricks also contains building blocks such as the one illustrated above, a 'rectangular torus', homeomorphic to a smooth torus. If we take a limiting process as before, the eight outer corners must still contribute $1/4$ to the Euler number, from which we deduce that the eight inner corners contribute $-1/4$, since the total Euler number is 0. For each $g > 0$, our child may construct a 'rectangular' g-holed torus, just by putting g of these blocks together in a line. There will still be eight outer corners, with a total contribution of 2 to the Euler number, but each hole now has a contribution of -2. Thus the total Euler number is $2 - 2g$, as expected.

We now elucidate the mathematics behind the calculation we have just performed. Let us consider a general polyhedron X in $\mathbf{R}^3$. Here, we are not assuming, as is sometimes done, that the polyhedron is topologically a sphere — it may for instance look like the children's building block illustrated above. It is assumed to be bounded, and so is therefore compact. The faces of X are plane polygons, and together they form a polygonal decomposition of the space X. As usual, we define the Euler number by $e = F - E + V$, where $F = \#$ faces, $E = \#$ edges and $V = \#$ vertices. In a similar way to the procedure we adopted for the cube, we may approximate X by a smooth surface S, which is locally Euclidean except for small neighbourhoods corresponding to the vertices. Let us consider one of the vertices P of X, and ask how much curvature we should expect to accumulate there, in the limiting sense explained above.

Suppose that r faces, say $\Pi_1, \ldots, \Pi_r$, meet at P. For d small, we consider the r-gon R on X determined by the r points at distance d from P along the r edges through P, with the sides of the polygon being line segments (on the faces) joining adjacent points. By taking d small enough, we can ensure that R does not meet any of the other edges of X. Let us consider two adjacent sides of this polygon, say P_0P_1 on Π_1 and P_1P_2 on Π_2. If we have taken the Euclidean metric on the complement of the vertices in X, we can locally flatten out the edge of the polyhedron containing the line PP_1, obtaining plane isosceles triangles PP_0P_1 and PP_1P_2. If the face Π_i has an angle θ_i at P, then the base angles of these two isosceles triangles are $(\pi - \theta_1)/2$, respectively $(\pi - \theta_2)/2$, and so the r-gon R on X has an angle $\pi - \theta_1/2 - \theta_2/2$ at its vertex P_1.

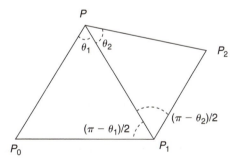

This argument generalizes to give the same fact for approximations of X by a smooth surface S with metric chosen so that the simple closed polygonal curve $P_1P_2 \ldots P_rP_1$ is contained in the open subset of S where the metric is locally Euclidean. Intuitively, if one thinks of the surface of X locally made out of folded paper, then for S we are locally bending the paper rather than folding it, and the edges of R still represent geodesic line segments (on S) meeting at the same angle as before.

If we let R' denote the corresponding r-gon on S, all this suggests that the contribution to the curvature integral from the relevant neighbourhood on S should be

$$\int_{R'} K \, dA = \sum \text{interior angles} \; - \; (r-2)\pi$$

$$= \sum_{i=1}^{r} (\pi - \theta_i) - (r-2)\pi = 2\pi - (\theta_1 + \cdots + \theta_r),$$

where here we have assumed the fact that the formula from Corollary 8.14 holds for R'.

Definition 8.18   The number $2\pi - (\theta_1 + \cdots + \theta_r)$ is called the *spherical defect* of the polyhedron X at the vertex P, and we shall denote it by $\text{defect}(P)$.

Since the curvature K is zero on the complement in S of these polygons, each corresponding to a vertex of X, the Gauss–Bonnet theorem applied to S therefore *suggests* that a corresponding discrete version should hold for any polyhedron X, with the Euler number of X being determined by these local contributions. This is indeed the case, and both the statement and proof are remarkably simple.

Proposition 8.19 (Discrete Gauss–Bonnet theorem)   *Let X denote a compact polyhedron in $\mathbf{R}^3$, with Euler number $e(X) = F - E + V$, where $F = \#\,faces$, $E = \#\,edges$ and $V = \#\,vertices$. If the vertices of X are $P_1, \ldots, P_V$ say, then*

$$\sum_{i=1}^{V} \text{defect}(P_i) = 2\pi \, e(X).$$

Proof   We denote the faces of X by $\Pi_1, \ldots, \Pi_F$. The sum of the spherical defects is then

$$2\pi V \; - \; \sum_{j=1}^{F} (\text{sum of the angles in } \Pi_j).$$

If Π_j is an m_j-gon for $1 \le j \le F$, this may be rewritten as

$$2\pi V \; - \; \sum_{j=1}^{F} (m_j - 2)\pi,$$

by the Euclidean Gauss–Bonnet formula for plane polygons. Since $\sum_{j=1}^{F} m_j = 2E$, each edge being the side of exactly two faces, the above formula reduces to $2\pi \, e(X)$.

$\square$

Exercises

8.1 For T the locally Euclidean torus, consider two charts obtained by projecting two different open unit squares from $\mathbf{R}^2$. Show that the corresponding transition function is not in general a translation, although it is *locally* a translation. What is the minimum number of such charts needed to form an atlas?

8.2 Verify, by explicit calculation, the global Gauss–Bonnet theorem for the embedded torus.

8.3 If $S \subset \mathbf{R}^3$ is a closed embedded surface with non-positive Euler number, deduce that there are points on S at which the curvature is positive, negative and zero.

8.4 Let P be a point on a smooth surface S, equipped with a Riemannian metric. Suppose that P has a normal neighbourhood W, with the property that, with respect to the corresponding geodesic polar coordinates (ρ, θ), the metric takes the form $d\rho^2 + f(\rho)^2 d\theta^2$, with $f = \sin\rho$, $f = \rho$ or $f = \sinh\rho$. Show that W is isometric to an open subset of, respectively, the sphere, the Euclidean plane, or the hyperbolic plane.

8.5 Suppose we have a Riemannian metric of the form $|dz|^2/h(r)^2$ on an open disc of radius $\delta > 0$ centred on the origin in $\mathbf{C}$ (possibly all of $\mathbf{C}$), where $h(r) > 0$ for all $r < \delta$. Show that the curvature K of this metric is given on the punctured disc by the formula
$$K = hh'' - (h')^2 + r^{-1}hh'.$$

8.6 Show that the embedded surface S with equation $x^2 + y^2 + c^2z^2 = 1$, where $c > 0$, is homeomorphic to the sphere. Deduce from the Gauss–Bonnet theorem that
$$\int_0^1 (1 + (c^2 - 1)u^2)^{-3/2} du = c^{-1}.$$

8.7 Let $S \subset \mathbf{R}^3$ be the catenoid, i.e. the surface of revolution corresponding to the curve $\eta(u) = (c^{-1}\cosh(cu), 0, u)$, for $-\infty < u < \infty$, where c is a positive constant. Show that S has infinite area, but that $\int_S K \, dA = -4\pi$.

8.8 Let $S \subset \mathbf{R}^3$ be the embedded surface given as the image of the open unit disc in $\mathbf{R}^2$ under the smooth parametrization
$$\sigma(u, v) = (u, v, \log(1 - u^2 - v^2))$$
— this may be thought of as obtained from a standard unit hemisphere by suitably stretching off to infinity in the negative z-direction. Verify that $\int_S K \, dA = 2\pi$.

8.9 Prove from first principles that a polyhedron in $\mathbf{R}^3$ must have at least one vertex where the spherical defect is positive. How is this result related to Proposition 6.19?

8.10 Given a topological triangle $\triangle$ with geodesic sides on a surface S (equipped with a Riemannian metric), and given $\varepsilon > 0$, show that there exists a polygonal decomposition of $\triangle$ whose polygons have diameters less than ε. Verify that the Euler number of such a polygonal decomposition is 1.

8.11 Using the previous exercise, together with Proposition 8.12 and Corollary 8.14, prove that the formula from Proposition 8.7 is valid for any topological triangle with geodesic sides on a surface S.

8.12   For $a > 0$, let S be the circular half-cone in $\mathbf{R}^3$ defined by $z^2 = a(x^2 + y^2)$, $z > 0$. Using the previous exercise, or otherwise, show that the curvature concentrated at the vertex (in the sense of Section 8.5) is given by the formula

$$2\pi(1 - (a + 1)^{-1/2}).$$

Postscript

We have now reached the end of this short course on Geometry. We have touched on some non-trivial mathematics, but we have done so in an explicit way, avoiding for the most part any general theories. The reader who has understood the material presented should be not only well informed on some important classical geometry, but also well prepared to take on these more general theories, which at a university in the UK might be taught in the third or fourth years. Examples of some of these standard theories are the following.

- Riemann surfaces: Here, local *complex* structures are put on our smooth surfaces. Our treatment of the hyperbolic plane is closely linked to the theory of *uniformization* of Riemann surfaces.
- Differential manifolds: Our treatment of abstract surfaces leads in higher dimensions to the study of differential manifolds and their properties.
- Algebraic topology: Our discussion of the Euler number and its topological invariance should motivate the development of homology groups of topological spaces.
- Riemannian geometry: Our treatment of Riemannian metrics, geodesics and curvature generalizes in a natural way to arbitrary dimensions, where the curvature of a Riemannian manifold is determined by the *sectional curvatures*, which, at any given point, are the Gaussian curvatures of two-dimensional sections (these corresponding via geodesics to the tangent planes at the point). The theory of these higher-dimensional *curved spaces* is of crucial importance to large areas of mathematics and theoretical physics.

References

[1]   A. F. Beardon *Complex Analysis: The Argument Principle in Analysis and Topology.* Chichester, New York, Brisbane, Toronto: Wiley, 1979.

[2]   A. F. Beardon *The Geometry of Discrete Groups.* New York, Heidelberg, Berlin: Springer-Verlag, 1983.

[3]   G. E. Bredon *Topology and Geometry.* New York, Heidelberg, Berlin: Springer-Verlag, 1997.

[4]   H. S. M. Coxeter *Introduction to Geometry.* New York: Wiley, 1961.

[5]   M. Do Carmo *Differential Geometry of Curves and Surfaces.* Englewood Cliffs, NJ: Prentice-Hall, Inc., 1976.

[6]   Jürgen Jost *Compact Riemann Surfaces: An Introduction to Contemporary Mathematics.* Berlin, Heidelberg, New York: Universitext, Springer-Verlag, 2002.

[7]   W. S. Massey *A Basic Course in Algebraic Topology.* New York, Heidelberg, Berlin: Springer-Verlag, 1991.

[8]   John McCleary *Geometry from a Differential Viewpoint.* Cambridge: Cambridge University Press, 1994.

[9]   A. Pressley *Elementary Differential Geometry.* Springer Undergraduate Mathematics Series, London: Springer-Verlag, 2001.

[10] Miles Reid and Balázs Szendrői *Geometry and Topology.* Cambridge: Cambridge University Press, 2005.

[11] W. Rudin *Principles of Mathematical Analysis.* New York: McGraw–Hill, 1976.

[12] M. Spivak *Differential Geometry, Volume 1.* Houston, TX: Publish or Perish, 1999.

[13] W. A. Sutherland *Introduction to Metric and Topological Spaces.* Oxford: Clarendon, 1975.

Index